Skills Worksheet
Directed Reading

Section: How Did Life Begin?

Complete each statement by writing the correct term or phrase in the space provided.

1. Scientists use _____ _____ to calculate the age of an object by measuring the proportions of the radioactive isotopes of certain elements.

2. Unstable isotopes that slowly change and give off energy in the form of charged particles are called _____ .

3. A radioisotope's _____ _____ is the time it takes for one-half of a given amount of a radioisotope to change.

Read each question, and write your answer in the space provided.

4. What is the primordial soup model?

   _____

   _____

5. Explain why the results of the Miller-Urey experiment have recently been reevaluated.

   _____

   _____

   _____

6. What is the bubble model?

   _____

   _____

In the space provided, write the letter of the term or phrase that best completes the statement.

_____ 7. RNA molecules can catalyze
   a. primordial soup.
   b. protein synthesis.
   c. ultraviolet radiation.
   d. coacervate development.

Copyright © by Holt, Rinehart and Winston. All rights reserved.

Holt Biology — History of Life on Earth

Name _____ Class _____ Date _____

Directed Reading *continued*

_____ 8. It is possible that cellular life began with
   a. oil and vinegar.
   b. water and lightning.
   c. large organic molecules.
   d. microspheres made of amino acids.

_____ 9. Microspheres could not be considered true cells unless they could
   a. form cellular membranes.
   b. originate spontaneously in water.
   c. incorporate molecules and energy.
   d. pass their characteristics to offspring.

_____ 10. Most scientists agree that double-stranded DNA evolved
   a. after RNA.
   b. before RNA.
   c. before microspheres.
   d. after hereditary mechanisms.

_____ 11. Most scientists agree that RNA first formed
   a. inside ammonia.
   b. outside the atmosphere.
   c. spontaneously in water.
   d. gradually in microspheres.

_____ 12. The simple chemical reactions on the early Earth were energized by
   a. enzymes.
   b. organic molecules.
   c. the sun and volcanoes.
   d. lightning.

Name _____ Class _____ Date _____

Skills Worksheet
Directed Reading

Section: The Evolution of Cellular Life

Read each question, and write your answer in the space provided.

1. What are the oldest fossils, and what have they told us about the first organisms?

   _____

   _____

2. How do eubacteria and archaebacteria differ?

   _____

   _____

Complete each statement by writing the correct term or phrase in the space provided.

3. The first eukaryotic cells are more likely to have evolved from _____ than from eubacteria.

4. The group _____ includes many bacteria that cause disease and decay.

Read each question, and write your answer in the space provided.

5. What is the theory of endosymbiosis?

   _____

   _____

6. What are chloroplasts?

   _____

   _____

   _____

Copyright © by Holt, Rinehart and Winston. All rights reserved.

Holt Biology — History of Life on Earth

Name _____ Class _____ Date _____

Directed Reading *continued*

In the space provided, write the letter of the term or phrase that best completes each statement.

_____ 7. The kingdom Protista consists of
   a. only unicellular prokaryotes.
   b. only multicellular eukaryotes.
   c. multicellular and unicellular prokaryotes.
   d. multicellular and unicellular eukaryotes.

_____ 8. An advantage of multicellularity is that
   a. cells can contain genetic material.
   b. organisms can live in many types of environments.
   c. cells can specialize to carry out specialized functions.
   d. organisms can be less complex than unicellular organisms.

_____ 9. Seaweed is classified as a
   a. plant.
   b. fungi.
   c. protist.
   d. prokaryote.

_____ 10. The kingdoms Fungi, Plantae, and Animalia each evolved independently from
   a. a single kind of protistan ancestor.
   b. a different kind of protistan ancestor.
   c. a single-celled prokaryote.
   d. a multicellular prokaryote.

_____ 11. The oldest known fossils of multicellular organisms were found in rocks
   a. older than the Cambrian period.
   b. formed in the Silurian period.
   c. formed in the early Precambrian era.
   d. younger than the Ordovician period.

Read each question, and write your answer in the space provided.

12. Why have mass extinctions changed the evolution of surviving species?

_____

_____

13. Why do some scientists think that another mass extinction is occurring today?

_____

_____

Copyright © by Holt, Rinehart and Winston. All rights reserved.

Holt Biology — History of Life on Earth

Name _____ Class _____ Date _____

Skills Worksheet

Directed Reading

Section: Life Invaded the Land

Study the following steps in the evolution of life on land. Determine the order in which the steps took place. Write the number of each step in the space provided.

_____ 1. The sun's rays caused some molecules of oxygen, O_2, to form molecules of ozone, O_3, in the upper atmosphere.

_____ 2. There were no living things on the dry, rocky surface of Earth.

_____ 3. Photosynthesis by cyanobacteria began adding oxygen to Earth's atmosphere.

_____ 4. Enough ozone had accumulated to make Earth's land a safe place to live.

_____ 5. In the upper atmosphere, ozone blocked the ultraviolet radiation of the sun.

Complete each statement by writing the correct term or phrase in the space provided.

6. Plants use the energy from sunlight to make their own _____ .

7. Plants cannot obtain _____ from bare rock.

8. Fungi cannot make _____ from sunlight.

9. Fungi are able to absorb _____ from bare rock.

10. Associations between fungi and the roots of plants are called _____ .

11. A relationship in which both organisms benefit is called _____ .

Read each question, and write your answer in the space provided.

12. Why was it necessary for plant life on land to evolve before animal life on land?

_____

_____

Name _____ Class _____ Date _____

Directed Reading *continued*

13. Describe the features of arthropods.

_____

_____

Complete each statement by writing the correct term or phrase in the space provided.

14. Early amphibians had moist breathing sacs called _____ .

15. Frogs, toads, and salamanders are examples of _____ .

16. Snakes, lizards, turtles, dinosaurs, and crocodiles are examples of _____ .

17. The movement of Earth's land masses over geologic time is commonly called _____ _____ .

Determine the order in which the following groups of animals evolved. Write the number of each step (1–4) in the space provided.

_____ **18.** amphibians

_____ **19.** mammals and birds

_____ **20.** fishes

_____ **21.** reptiles

Name _____ Class _____ Date _____

Skills Worksheet

Active Reading

Section: How Did Life Begin?

Read the passage below. Then answer the questions that follow.

Listed below are the steps of Louis Lerman's bubble model.

Step 1: Eruption of undersea volcanoes produces ammonia, methane, and other gases that become trapped in underwater bubbles.

Step 2: Protected by bubbles, gases needed to make amino acids undergo chemical reactions.

Step 3: As bubbles burst on the water's surface, simple organic molecules are released into the air.

Step 4: The simple organic molecules are carried upward by the wind and exposed to ultraviolet radiation and lightning. The additional energy they produce causes further reactions.

Step 5: Complex organic molecules fall into the oceans.

SKILL: ORGANIZING INFORMATION

Read each question, and write your answer in the space provided.

1. What gases were produced by the eruption of undersea volcanoes?

   _____

2. What were the simple organic molecules exposed to, as they were carried upward by wind?

   _____

3. After undergoing further reactions caused by exposure to ultraviolet radiation and lightning, what did organic molecules form?

   _____

In the space provided, write the letter of the phrase that best completes the statement.

_____ 4. According to Lerman's bubble model, the key processes that formed the chemicals needed for life took place
   a. at the same rate as the primordial soup model.
   b. more slowly than what is estimated by the primordial soup model.
   c. within bubbles on the ocean's surface.
   d. on land rather than on a watery surface.

Name _____ Class _____ Date _____

Skills Worksheet
Active Reading

Section: Complex Organisms Developed
Read the passage below. Then answer the questions that follow.

Early in the history of life, two different groups of prokaryotes evolved—eubacteria and archaebacteria. Living examples include *Escherichia coli*, a species of eubacteria, and *Sulfolobus*, a group of archaebacteria. Eubacteria are prokaryotes that contain a chemical called peptidoglycan in their cell walls and have the same type of lipids in their cell membranes that eukaryotes do. Eubacteria include many bacteria that cause disease and decay.

Archaebacteria are prokaryotes that lack peptidoglycan in their cell walls and have unique lipids in their cell membranes. Archaebacteria are thought to be closely related to the first bacteria to have existed on Earth. Chemical evidence indicates that the first eukaryotic cells are more likely to have evolved from archaebacteria than from eubacteria.

SKILL: READING EFFECTIVELY

Read each question, and write your answer in the space provided.

1. How are eubacteria and archaebacteria alike?

2. What relationship exists between *Sulfolobus* and archaebacteria?

3. What two traits of eubacteria are identified in the third sentence of this passage?

4. What two traits of archaebacteria are identified in the fifth sentence of this passage?

5. What unique trait of archaebacteria is described in the sixth sentence?

Active Reading *continued*

6. What evidence supports this idea?

_____

_____

An analogy is a comparison. In the space provided, write the letter of the term or phrase that best completes the analogy.

_____ **7.** Archaebacteria is to *Sulfolobus* as eubacteria is to
   a. *Escherichia coli.*
   b. cyanobacteria.
   c. peptidoglycan.
   d. Both (a) and (b)

Name _____ Class _____ Date _____

Skills Worksheet

Active Reading

Section: Life Invaded the Land

Read the passage below. Then answer the questions that follow.

The first vertebrates to inhabit the land were early amphibians. Amphibians are smooth-skinned organisms that include frogs, toads, and salamanders.

Amphibians were able to adapt to land because of the development of several structural changes in their bodies. Early amphibians had moist breathing sacs called lungs, which they used to absorb oxygen from air. The limbs of amphibians are thought to be derived from the bones of fish fins. What made walking possible was the evolution of a strong support system of bones in the region just behind the head. This system of bones provided a rigid base for the limbs to work against.

SKILL: READING EFFECTIVELY

Read each question, and write your answer in the space provided.

1. What three types of amphibians are identified in the passage?

   _____

   _____

2. What made it possible for amphibians to adapt to life on land?

   _____

   _____

3. According to the passage, the limbs of an amphibian were derived from what structure?

   _____

   _____

In the space provided, write the letter of the phrase that best completes the statement.

_____ 4. Jawless fishes and salamanders are alike in that both types of animals
   a. are amphibians.
   b. have backbones.
   c. have fins.
   d. are smooth-skinned organisms.

Name _____ Class _____ Date _____

Skills Worksheet
Vocabulary Review

In the space provided, explain how the terms in each pair differ in meaning.

1. radioisotope, half-life

   _____
   _____
   _____
   _____

2. fossil, cyanobacteria

   _____
   _____
   _____
   _____

3. eubacteria, archaebacteria

   _____
   _____
   _____
   _____

4. mycorrhiza, mutualism

   _____
   _____
   _____
   _____

Copyright © by Holt, Rinehart and Winston. All rights reserved.

Holt Biology   13   History of Life on Earth

Name _____ Class _____ Date _____

Vocabulary Review *continued*

In the space provided, write the letter of the description that best matches the term or phrase.

_____ 5. radiometric dating

_____ 6. endosymbiosis

_____ 7. protists

_____ 8. mass extinction

_____ 9. arthropod

_____ 10. vertebrate

_____ 11. continental drift

_____ 12. microsphere

a. animal with hard outer skeleton and jointed limbs

b. the movement of Earth's land masses over geologic time

c. animals with backbones

d. calculation of the age of an object by measuring the proportions of radioactive isotopes of certain elements

e. tiny droplets made of short chains of amino acids

f. the theory that mitochondria and chloroplasts are the descendants of symbiotic aerobic eubacteria

g. members of a kingdom of unicellular and multicellular eukaryotic organisms

h. the death of all members of many different species

Name _____ Class _____ Date _____

Skills Worksheet
Science Skills

Interpreting Timelines

The timeline below shows some of the physical events that have helped to shape life on Earth. Some major advances in the evolution of life are also shown. Use the timeline and the "Life Events" (*a–d*) listed below the timeline to complete items 1–4.

In each numbered space in the timeline, write the letter of the most appropriate event from the "Life Events" below the timeline.

Timeline of Life on Earth

Number of years ago	Life events	Physical events
4.5 billion		Earth forms and surface cools.
2.5 billion	1. _____	Oxygen gas is released into seas, then enters atmosphere.
1.5 billion	The first eukaryotes evolve.	
700 million	The first multicellular organisms evolve.	
440 million	First mass extinction	
430 million	2. _____	Protective ozone shield in place in upper atmosphere
370 million	Amphibians are the first vertebrates on land.	
360 million	Second mass extinction	
350 million	3. _____	Widespread drought
245 million	Third mass extinction	
210 million	Fourth mass extinction	
65 million	Fifth mass extinction	
60 million	4. _____	Earth's climate becomes moist.
2 million	The first humans appear.	
Present	More than half of tropical rain forests and many species destroyed	

Life Events
a. Reptiles evolve from amphibians and become dominant on land.
b. Birds and mammals become dominant on land.
c. Plants and fungi invade land for the first time.
d. Cyanobacteria first begin to carry out photosynthesis.

Name _____ Class _____ Date _____

Science Skills continued

Use the timeline on the previous page to answer questions 5–9.

Read each question, and write your answer in the space provided.

5. Name an organism that is seriously damaging Earth's ecosystems, and describe some of the damage being caused by this organism.

   _____
   _____
   _____

6. What are some factors that have caused the dominant life-forms on Earth to change over time?

   _____
   _____
   _____
   _____
   _____

7. Is another major mass extinction likely? Explain.

   _____
   _____
   _____

8. What environmental changes affected the evolution of reptiles?

   _____
   _____
   _____

9. What effect did photosynthetic cyanobacteria have on Earth's environment?

   _____
   _____
   _____

Name _____ Class _____ Date _____

Skills Worksheet

Concept Mapping

Using the terms and phrases provided below, complete the concept map showing the evolution of early life-forms.

chloroplasts   fossils   prokaryotes
cyanobacteria   life on land   2.5 billion years ago
endosymbiosis   mitochondria   ozone
eubacteria

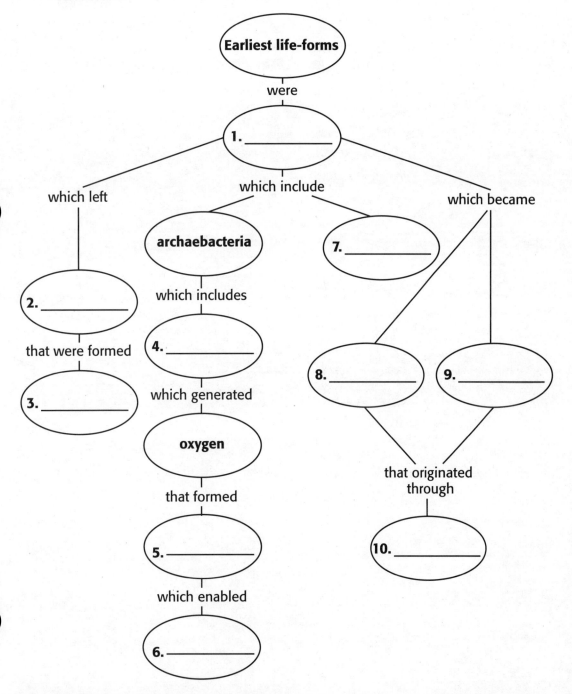

Name _____ Class _____ Date _____

Skills Worksheet
Critical Thinking

Work-Alikes

In the space provided, write the letter of the term or phrase that best describes how each numbered item functions.

_____ 1. radiometric dating

_____ 2. RNA

_____ 3. proteins, carbohydrates, lipids, and nucleic acids

_____ 4. cyanobacteria

_____ 5. multicellular organism

_____ 6. mass extinction

_____ 7. ozone

a. great shield

b. team of workers

c. oxygen tank

d. entire aquarium of fish die

e. building blocks

f. catalyst

g. ticking of a clock

Cause and Effect

In the space provided, write the letter of the term or phrase that best matches each cause or effect given below.

Cause	Effect	
8. _____	chemical reactions proceeded faster	**a.** they evolved partnerships called mycorrihizae
9. bacteria entered large cells	_____	**b.** development of amphibian lung
10. evolution of wings	_____	**c.** insects could patrol the landscape
11. _____	oxygen obtained from the air	**d.** bubbles formed
12. fungi absorb the minerals used by the plant roots that the fungi live on	_____	**e.** they evolved into mitochondria and chloroplasts
13. Earth has a protective layer of ozone gas	_____	**f.** its surface is protected from ultraviolet radiation

Copyright © by Holt, Rinehart and Winston. All rights reserved.

Holt Biology — History of Life on Earth

Name _____ Class _____ Date _____

Critical Thinking continued

Linkages

In the spaces provided, write the letters of the two terms or phrases that are linked together by the term or phrase in the middle. The choices can be placed in any order. Some choices may be used more than once.

14. _____ radiometric dating _____
15. _____ catalysis _____
16. _____ tropical rain forests _____
17. _____ vertebrates on land _____
18. _____ eggs laid on land _____
19. _____ current positions of continents _____

a. lungs and enhanced circulation
b. proteins formed
c. development of limbs
d. continental drift
e. initial RNA molecules
f. destruction of rain forests
g. marsupials found in Australia and South America
h. age of rock estimated
i. human activity
j. development of watertight eggs
k. half-life
l. terrestrial reptiles

Analogies

An analogy is a relationship between two pairs of terms or phrases written as a : b :: c : d. The symbol : is read as "is to," and the symbol :: is read as "as." In the space provided, write the letter of the pair of terms or phrases that best completes each analogy shown.

_____ 20. simple organic molecules : ocean bubbles ::
   a. oceans : bubbles
   b. cells : microspheres
   c. microspheres : short chains of amino acids
   d. microspheres : oil mixed with water

_____ 21. bubble model : primordial soup model ::
   a. skate board : automobile
   b. runner : walker
   c. hummingbird : eagle
   d. turtle : cheetah

_____ 22. bacteria : prokaryotic ::
   a. protists : eukaryotes
   b. eukaryotes : prokaryotes
   c. cyanobacteria : eukaryotes
   d. eubacteria : eukaryotes

_____ 23. eukaryotic cells : archaebacteria ::
   a. bacteria : arthropods
   b. jawed fishes : jawless fishes
   c. plants : insects
   d. fish : mammals

_____ 24. ozone : ultraviolet radiation ::
   a. carbon dioxide : oxygen
   b. nitrogen : protons
   c. methane : microspheres
   d. umbrella : rain

Name _____ Class _____ Date _____

Skills Worksheet
Test Prep Pretest

In the space provided, write the letter of the term or phrase that best completes each statement or best answers each question.

_____ 1. Radiometric dating has determined that Earth is approximately
   a. 4,000 years old.
   b. 500,000 years old.
   c. 2.5 billion years old.
   d. 4.5 billion years old.

_____ 2. A mechanism for heredity was necessary in order to begin
   a. microspheres.
   b. life.
   c. RNA.
   d. protein.

_____ 3. The kingdoms that evolved from protists are
   a. bacteria, fungi, and animals.
   b. bacteria, plants, and animals.
   c. plants, animals, and humans.
   d. fungi, plants, and animals.

_____ 4. Life was able to move from the sea to land because
   a. photosynthesis by cyanobacteria added oxygen to Earth's atmosphere.
   b. ozone was created from the oxygen produced by photosynthesis.
   c. ozone provides a shield from the harsh ultraviolet rays of the sun.
   d. All of the above

_____ 5. Louis Lerman's bubble model addresses the problem of
   a. lightning.
   b. ultraviolet radiation damage to ammonia and methane.
   c. volcanic heat.
   d. a dense ozone layer.

_____ 6. The following animals are all arthropods.
   a. crabs, lobsters, insects, and spiders
   b. crabs, snakes, insects, and spiders
   c. frogs, toads, and salamanders
   d. frogs, toads, salamanders, and snakes

_____ 7. The primordial soup model requires
   a. the sun, electrical energy, or volcanic eruptions.
   b. at least 1 billion years.
   c. hydrogen-containing gases.
   d. All of the above

Name _____ Class _____ Date _____

Test Prep Pretest *continued*

_____ 8. Amphibians were able to adapt to life on land for all of the following reasons EXCEPT
  a. lungs.
  b. watertight skin.
  c. limbs.
  d. a platform of bone that provided a base for the limbs to work against.

_____ 9. Reptiles are more completely adapted to land than amphibians are because reptiles
  a. have watertight skin.
  b. can lay their eggs on dry land.
  c. do not have to live near the water.
  d. All of the above

_____ 10. Scientists think the first step toward cellular organization was
  a. nucleotides.   c. microspheres.
  b. coacervates.   d. RNA enzymes.

In the space provided, write the letter of the description that best matches the term or phrase.

_____ 11. bacteria

_____ 12. mass extinctions

_____ 13. plants and fungi

_____ 14. RNA

_____ 15. jawless fishes

_____ 16. endosymbiosis

_____ 17. multicellularity

_____ 18. continental drift

a. the first self-replicating information storage molecule

b. thought to have evolved 2.5 billion years ago

c. cause decreased competition for resources among survivors

d. the first multicellular organisms to live on land

e. enabled cell specialization

f. explains why related animals are found on continents separated by oceans

g. the first vertebrates

h. the theory that mitochondria and chloroplasts are descendants of symbiotic, aerobic eubacteria

Complete each statement by writing the correct term or phrase in the space provided.

19. In Louis Lerman's model, the hydrogen-containing gases needed to make amino acids were trapped in _____ _____ .

Name _____ Class _____ Date _____

Test Prep Pretest *continued*

20. The earliest stages of cellular organization may have been the formation of

    _____ .

21. Because of the mass extinction at the end of the Permian period, about

    96 percent of all species of _____ living at the time became extinct.

22. A partnership, such as the one found in mycorrhizae, in which each organism

    helps the other survive is called _____ .

23. The first group of animals to live on land was the _____ .

Read each question, and write your answer in the space provided.

24. Why was the development of multicellular organisms a great step forward in the evolution of life on Earth?

    _____

    _____

25. What were the first vertebrates to live on land? What structural changes enabled them to make the transition?

    _____

    _____

    _____

    _____

    _____

    _____

Name _____ Class _____ Date _____

Assessment Quiz

Section: How Did Life Begin?

In the space provided, write the letter of the term or phrase that best completes each statement or best answers each question.

_____ 1. Approximately how many half-lives of potassium-40 have passed since the formation of Earth?
   a. 2
   b. 3.5
   c. 4.5
   d. 5

_____ 2. Which mixture of gases is thought to have made up Earth's early atmosphere?
   a. nitrogen, hydrogen, water vapor, ammonia, methane
   b. carbon dioxide, nitrogen, oxygen, argon, ozone
   c. ozone, water vapor, carbon dioxide, ammonia
   d. methane, oxygen, hydrogen, argon, water vapor

_____ 3. Most scientists agree that the basic molecules of life could have formed spontaneously through simple chemistry on the early Earth. Which of the following has been made to form spontaneously in water in the laboratory?
   a. proteins
   b. DNA
   c. RNA
   d. None of the above

_____ 4. Microspheres could not be considered true cells unless they had the characteristics of living things, including
   a. intelligence.
   b. nuclei.
   c. coacervates.
   d. heredity.

In the space provided, write the letter of the description that best matches each term or phrase.

_____ 5. radioisotopes

_____ 6. primordial soup model

_____ 7. microspheres

_____ 8. RNA

_____ 9. radioactive decay

_____ 10. bubble model

a. a measurable process of physics that enables scientists to estimate the age of ancient objects

b. flawed by not accounting for the lack of ozone in the upper atmosphere

c. thought to represent how organic molecules could have developed quickly, protected from ultraviolet radiation

d. thought to have been the precursors to living cells

e. thought to have catalyzed the assembly of the first proteins

f. unstable isotopes that break down and give off energy

Assessment Quiz

Section: The Evolution of Cellular Life

In the space provided, write the letter of the term or phrase that best completes each statement or best answers each question.

_____ 1. Which group of prokaryotes includes many species that cause disease and decay?
   a. eubacteria
   b. peptidoglycans
   c. cyanobacteria
   d. archaebacteria

_____ 2. The first eukaryotic kingdom containing both multicellular and single-celled organisms is
   a. Eubacteria.
   b. Protista.
   c. Archaebacteria.
   d. Animalia.

_____ 3. Which is one advantage of multicellularity?
   a. The organisms can protect themselves from predators.
   b. Cells can dry out faster.
   c. The organisms can find mates better.
   d. Cells can carry out specialized functions.

_____ 4. Some mass extinctions occur due to natural causes. The one that is thought to be occurring now is happening because of
   a. climate changes.
   b. human activity.
   c. meteorite impacts.
   d. geological changes.

_____ 5. The invading bacteria that became chloroplasts through the process of endosymbiosis were probably closely related to
   a. eubacteria.
   b. *Escherichia coli.*
   c. cyanobacteria.
   d. *Sulfolobus.*

In the space provided, write the letter of the event that best matches each time.

_____ 6. 2.5 billion years ago   a. diverse protists

_____ 7. 1.5 billion years ago   b. earliest multicellular organisms

_____ 8. 1 billion years ago   c. earliest fossil bacteria

_____ 9. 700 million years ago   d. first mass extinction

_____ 10. 440 million years ago   e. first eukaryotes

Name _____ Class _____ Date _____

Assessment
Quiz

Section: Life Invaded the Land

In the space provided, write the letter of the term or phrase that best completes each statement or best answers each question.

_____ 1. Oxygen began to enter the atmosphere of the early Earth as a result of
  a. respiration by microspheres.
  b. bubbles bursting at the surface of the oceans.
  c. photosynthesis by cyanobacteria.
  d. endosymbiosis of prokaryotes and eukaryotes.

_____ 2. For which reason did multicellular fungi and plants leave the oceans together to live on land?
  a. They each possessed a characteristic needed by the other.
  b. Fungi are predators of plants.
  c. Plants needed fungi for pollination.
  d. They could only produce oxygen if they were together.

_____ 3. Which of the following enabled arthropods to inhabit land?
  a. Arthropods could not live on land until wings had evolved.
  b. Flowering plants had to evolve first.
  c. The climate dried out enough for scorpion-like arthropods to leave the oceans.
  d. Plants had already covered Earth and provided a food source for arthropods.

_____ 4. Which major structural change had to be made in body organization before descendants of fishes were capable of living on land?
  a. wings   c. lungs
  b. jaws   d. backbone

_____ 5. Which statement best explains why there are large numbers of marsupials in Australia and South America and few anywhere else on Earth?
  a. Marsupials evolved independently in several locations.
  b. Early marsupials evolved in one location and became separated by continental drift.
  c. Marsupials can live only in the Australian and South American climates.
  d. Marsupials crossed the oceans from Australia to South America on floating debris.

Holt Biology — History of Life on Earth

Name _____ Class _____ Date _____

Quiz *continued*

In the space provided, write the letter of the description that best matches the term or phrase.

_____ **6.** mycorrhizae

_____ **7.** mutualism

_____ **8.** vertebrate

_____ **9.** arthropod

_____ **10.** amphibian

a. a kind of animal with a hard outer skeleton, a segmented body, and jointed limbs

b. an animal with a backbone

c. symbiotic relationship between two species in which both benefit

d. smooth-skinned animal possessing moist breathing sacs

e. symbiotic associations between fungi and plant roots

Name _____ Class _____ Date _____

Assessment
Chapter Test

History of Life on Earth

In the space provided, write the letter of the description that best matches the term or phrase.

_____ 1. protists

_____ 2. ozone

_____ 3. insects

_____ 4. ammonia

_____ 5. reptiles

a. most numerous group of animals

b. first to lay eggs with shells

c. required by primordial soup model

d. the first eukaryotes

e. absorbs ultraviolet radiation

In the space provided, write the letter of the term or phrase that best completes each statement or best answers each question.

_____ 6. Cyanobacteria changed the early Earth's atmosphere by giving off
   a. carbon dioxide.
   b. ammonia.
   c. hydrogen.
   d. oxygen.

_____ 7. All the living things on Earth today are grouped into
   a. three kingdoms.
   b. four kingdoms.
   c. five kingdoms.
   d. six kingdoms.

_____ 8. Eukaryotes may have descended from
   a. eubacteria.
   b. archaebacteria.
   c. cyanobacteria.
   d. None of the above

_____ 9. Determining the age of a rock by comparing relative proportions of its radioactive isotopes is called
   a. radioactive decay.
   b. radiometric dating.
   c. isotope dating.
   d. half-lives.

_____ 10. Protists were the first organisms to exhibit cell specialization, which was only possible when they
   a. grew large enough.
   b. became eukaryotic.
   c. evolved into fungi, plants, and animals.
   d. became multicellular.

_____ 11. The first step toward cellular organization may have been the gathering of amino acids into
   a. macromolecules.
   b. coacervates.
   c. microspheres.
   d. DNA molecules.

Copyright © by Holt, Rinehart and Winston. All rights reserved.

Holt Biology

Chapter Test *continued*

_____ 12. Organisms were able to live safely on dry land after
   a. cyanobacteria made oxygen, and ozone began to form.
   b. the fifth mass extinction.
   c. continental drift stopped.
   d. archaebacteria made oxygen, and ozone began to form.

_____ 13. Which of the following enabled the formation of true cells?
   a. the ozone layer
   b. heredity
   c. microspheres
   d. endosymbiosis

_____ 14. The first animals to live successfully on land were
   a. amphibians.
   b. reptiles.
   c. arthropods.
   d. flying insects.

_____ 15. In mycorrhizae, fungi provide plants with
   a. food.
   b. energy.
   c. minerals.
   d. water.

_____ 16. Prokaryotes that lack peptidoglycan in their cell walls are called
   a. eubacteria.
   b. *Escherichia coli*.
   c. pre-eukaryotes.
   d. archaebacteria.

_____ 17. Each of Earth's great mass extinctions has been followed by
   a. the destruction of Earth's ecosystems.
   b. a burst of evolution.
   c. geological and weather changes.
   d. overpopulation of bacteria.

_____ 18. Fungi, plants, and animals are
   a. organisms that live only on land.
   b. organisms that form mycorrhizae.
   c. three successful kingdoms that evolved from protists.
   d. three successful kingdoms that evolved from eubacteria.

_____ 19. The most successful living vertebrates are
   a. humans.
   b. insects.
   c. reptiles.
   d. fishes.

_____ 20. Insects probably owe their successful numbers and diversity to the evolution of
   a. legs.
   b. wings.
   c. vision.
   d. flowers.

Name _____ Class _____ Date _____

Assessment

Chapter Test

History of Life on Earth

In the space provided, write the letter of the term or phrase that best completes each statement or best answers each question.

____ 1. Radiometric dating compares the proportions of specific radioisotopes with their
    a. atomic mass.
    b. more stable isotopes.
    c. charged particles.
    d. unstable isotopes.

____ 2. In a unique biological association known as mycorrhizae, fungi provide
    a. oxygen to plants through photosynthesis.
    b. food to plants, and the plants provide minerals to the fungi.
    c. minerals to plants, and the plants provide food to the fungi.
    d. protein to plants through the fungal root system.

____ 3. The first vertebrates on land were
    a. reptiles.
    b. insects.
    c. amphibians.
    d. fishes.

____ 4. Primitive cyanobacteria
    a. were among the first bacteria to appear on Earth.
    b. produced the first oxygen in Earth's atmosphere.
    c. are believed to be the ancestors of chloroplasts.
    d. All of the above

____ 5. Reptiles evolved from
    a. birds.
    b. mammals.
    c. arthropods.
    d. amphibians.

____ 6. Birds and mammals have become the dominant vertebrates on land partly because of
    a. continental drift.
    b. the evolution of wings.
    c. the shift toward a moister climate.
    d. the evolution of a shelled egg.

____ 7. The earliest traces of life on Earth are fossils of
    a. trilobites.
    b. bacteria.
    c. dinosaurs.
    d. mycorrhizae.

____ 8. The formation of ozone in the upper atmosphere caused the Earth's surface to be
    a. hot, barren, and rocky.
    b. hazardous to living things.
    c. a safe place to live for the first time.
    d. flooded with dangerous ultraviolet radiation.

Name _____ Class _____ Date _____

Chapter Test *continued*

_____ 9. A newly formed rock containing potassium-40 will have which proportion of the original potassium-40 remaining after three half-lives?
  a. 1/2
  b. 1/4
  c. 1/8
  d. 1/16

_____ 10. The breakdown of unstable isotopes results in
  a. more reactive isotopes.
  b. smaller and more stable isotopes.
  c. elements of greater atomic mass.
  d. an increase in the half-life of the isotope.

_____ 11. Which observation helps support the idea that mitochondria and chloroplasts descended from bacteria?
  a. Both have circular DNA similar to that found in bacteria.
  b. Neither is capable of existing outside a living cell.
  c. Neither has the same organelles that aerobic eubacteria have.
  d. Both reproduce by mitosis.

_____ 12. In what way did the mass extinction of the dinosaurs 65 million years ago allow for the rise of mammals and birds?
  a. Many resources became available to the surviving mammals and birds.
  b. The climate became more moist, and the reptiles' advantage was not so great.
  c. Small, feathered reptiles survived and gave rise to modern birds.
  d. All of the above

_____ 13. The third and most devastating of all mass extinctions occurred at the end of which period of Earth's history?
  a. Triassic
  b. Permian
  c. Devonian
  d. Jurassic

_____ 14. What is the significance to the development of life on Earth of the first formation of RNA molecules?
  a. They were capable of storing information.
  b. They were self-replicating.
  c. They could change from one generation to the next.
  d. All of the above

_____ 15. Plants likely evolved from
  a. protists.
  b. fungi.
  c. eubacteria.
  d. mycorrhizae.

Chapter Test *continued*

In the space provided, write the letter of the description that best matches the term or phrase.

_____ 16. radiometric dating

_____ 17. the primordial soup model

_____ 18. the bubble model

_____ 19. heredity

_____ 20. prokaryotes

_____ 21. eukaryotes

_____ 22. Cambrian period

_____ 23. arthropods

_____ 24. vertebrates

_____ 25. continental drift

a. Over geologic time, this process contributed to the geographic distribution of some species.

b. Some examples are fishes, salamanders, dinosaurs, birds, and humans.

c. first organisms to have wings whose success is probably connected to the ability to fly

d. hypothesis that the early oceans were filled with many different organic molecules formed by chemical reactions energized by the sun, volcanoes, and lightning

e. appeared about 1.5 billion years ago and have a complex system of internal membranes

f. appeared about 2.5 billion years ago and form two groups, one of which contains organisms that cause disease

g. hypothesis that the processes that formed the chemicals needed for life took place within an environment protected from ultraviolet radiation

h. characterized by great evolutionary expansion and origination of most phyla that exist today

i. characteristic of living things that was initiated by the linking of amino acids and sugars within microspheres and coacervates, and the evolution of RNA molecules

j. method used to compute how many half-lives have passed since a rock was formed

Name _____ Class _____ Date _____

Quick Lab

DATASHEET FOR IN-TEXT LAB

Modeling Radioactive Decay

You can use some dried corn, a box, and a watch to make a model of radioactive decay that will show you how scientists measure the age of objects.

MATERIALS
- approximately 100 dry corn kernels per group
- cardboard box
- clock or watch with a second hand

Procedure

1. Assign one member of your team to keep time.
2. Place 100 dry corn kernels into a box.
3. Shake the box gently from side to side for 10 seconds.
4. Keep the box still and remove and count the kernels that "point" to the left side of the box. Record in the data table below the number of kernels you removed.

Data Table		
Total shake time (seconds)	Number of kernels removed	Number of kernels remaining
10		
20		
30		
40		
50		
60		

5. Repeat steps 4 and 5 until all kernels have been counted and removed.
6. Calculate the number of kernels remaining for each time interval.
7. Make a graph using your group's data. Plot "Total shake time (seconds)" on the x-axis. Plot "Number of kernels remaining" on the y-axis. Use a separate sheet of graph paper.

Modeling Radioactive Decay continued

Analysis

1. Identify what the removed kernels represent in each step.

_____

_____

2. Calculate the half-life of your sample, in seconds, that is represented in this activity.

3. Calculate the age of your sample, in years, if each 10-second interval represents 5,700 years.

4. Evaluate the ability of this model to demonstrate radioactive decay.

_____

_____

_____

_____

Name _____ Class _____ Date _____

Quick Lab

DATASHEET FOR IN-TEXT LAB

Modeling Coacervates

By using simple chemistry, you will see that some properties of coacervates resemble the properties of cells.

MATERIALS
- safety goggles and lab apron
- graduated cylinder
- 1 percent gelatin solution
- 1 percent gum arabic solution
- test tube
- 0.1 M HCl
- pipet
- microscope slide and coverslip
- microscope

Procedure

1. **CAUTION: Hydrochloric acid is corrosive. Put on safety goggles, gloves, and a lab apron. Avoid contact with skin and eyes. Avoid breathing vapors.** If any of this solution spills on you, immediately flush the area with water, and notify your teacher.

2. Mix 5 mL of a 1 percent gelatin solution with 3 mL of a 1 percent gum arabic solution in a test tube.

3. Add 0.1 M HCl to the gelatin-gum arabic solution one drop at a time until the solution turns cloudy.

4. Prepare a wet mount of the cloudy solution, and examine it under a microscope at high power.

5. Prepare a drawing of the structures that you see.

Name _____ Class _____ Date _____

Modeling Coacervates *continued*

Analysis

1. Describe what happened to the solutions after the acid was added.

_____

_____

2. Compare the appearance of coacervates with that of cells.

_____

_____

_____

3. Predict what would happen to the coacervates if a base was added to the solution.

_____

_____

_____

4. Critical Thinking
Evaluating Hypotheses Based on the evidence you obtained, defend the hypothesis that coacervates could have been the basis of life on Earth

_____

_____

_____

_____

Data Lab
Analyzing Signs of Endosymbiosis

DATASHEET FOR IN-TEXT LAB

Background

You may recall that mitochondria have their own DNA and produce their own proteins. The data below were collected by scientists studying the proteins produced by mitochondrial DNA. The scientists found that the three-nucleotide sequences (codons) in the nucleus of an organism's cells can code for different amino acids than those coded for in the cell's mitochondria. Examine the data below, and answer the questions that follow.

Amino Acids Made in the Nucleus and Mitochondria			
	Amino acids or other instructions coded for in the nucleus	Amino acids or other instructions coded for in mitochondria	
Codon	Plants and mammals	Plants	Mammals
UGA	Stop	Stop	Tryptophan
AGA	Arginine	Arginine	Stop
AUA	Isoleucine	Isoleucine	Methionine
AUU	Isoleucine	Isoleucine	Methionine
CUA	Leucine	Leucine	Leucine

Analysis

1. **Defend** the theory of endosymbiosis using these data.

2. **Infer** what these data indicate about the evolution of plant cells.

3. **Describe** how these data can be used to support the idea that more than one type of cell evolved early in the history of life.

Name _____ Class _____ Date _____

Exploration Lab

DATASHEET FOR IN-TEXT LAB

Making a Timeline of Life on Earth

SKILLS
- Observing
- Inferring relationships
- Organizing data

OBJECTIVES
- **Compare** and **contrast** the distinguishing characteristics of representative organisms of the six kingdoms.
- **Organize** the appearance of life on Earth in a timeline.

MATERIALS
- adding-machine tape (5 m roll)
- meterstick
- colored pens or pencils
- photographs or drawings of organisms from ancient Earth to present day

Before You Begin

About 4.5 billion years ago, Earth was a ball of molten rock. As the surface cooled, a rocky crust formed and water vapor in the atmosphere condensed to form rain. By 3.9 billion years ago, oceans covered much of Earth's surface. Rocks formed in these oceans contain **fossils** of bacterial cells that lived about 3.5 billion years ago. The **fossil record** shows a progression of life-forms and contains evidence of many changes in Earth's surface and atmosphere.

In this lab, you will make a timeline showing the major events in Earth's history and in the history of life on Earth, such as the evolution of new groups of organisms and the mass extinctions. This timeline can be used to study how living things have changed over time.

1. Write a definition for each boldface term in the paragraphs above. Use a separate sheet of paper.
2. Record your data in the data table provided.
3. Based on the objectives for this lab, write a question you would like to explore about the history of life on Earth.

_____

_____

Name _____ Class _____ Date _____

Making a Timeline of Life on Earth *continued*

Procedure

PART A: MAKING A TIMELINE

1. Make a mark every 20 cm along a 5 m length of adding-machine tape. Label one end of the tape "5 billion years ago" and the other end "Today." Write "20 cm = 200 million years" near the beginning of your timeline.

2. Locate and label a point representing the origin of Earth on your timeline. Use your textbook as a reference. See the timeline at the bottom of Section 2 and Section 3 of this chapter. Also locate and label the 11 periods of the geologic time scale beginning with the Cambrian period.

3. Using your textbook as a reference, mark the following events on your timeline: the first cyanobacteria appear; oxygen enters the atmosphere; the five mass extinctions; the first eukaryotes appear; the first multicellular organisms appear; the first vertebrates appear; the first plants, fungi, and land animals appear; the first dinosaurs and mammals appear; the first flowering plants appear; the first humans appear.

4. Look at the photographs of organisms provided by your teacher. Identify the major characteristics of each organism. Record your observations in the data table below.

Data Table		
Organism	**Kingdom**	**Characteristics/adaptation for life on Earth**

Name _____ Class _____ Date _____

Making a Timeline of Life on Earth *continued*

5. Lay out your timeline on the floor in your classroom. Place photographs (or drawings) of the organisms you examined on your timeline to show when they appeared on Earth.

6. Fold the timeline at the mark representing 4.8 billion years ago. This leaves 24 segments, each representing 200 million years on your timeline. Now you can think of each segment as 1 hour in a 24-hour day.

7. When you are finished, walk slowly along your timeline. Note the sequence of events in the history of life on Earth and the relative amount of time between each event.

PART B: CLEANUP AND DISPOSAL

8. Dispose of paper scraps in the designated waste container.

9. Clean up your work area and all lab equipment. Return lab equipment to its proper place.

ANALYZE AND CONCLUDE

1. **Analyzing Information** Think of each segment of your timeline as 1 hour in a 24-hour day as you answer each of the following questions.

    a. How long has life existed on Earth?

    _____

    b. For what part of the day did only unicellular life-forms exist?

    _____

    c. At what time of day did the first plants appear on Earth?

    _____

    d. At what time of day did mammals appear on Earth?

    _____

2. **Summarizing Information** Identify the major developments in life-forms that have occurred over the last 3.5 billion years.

    _____

    _____

    _____

    _____

Name _____ Class _____ Date _____

Making a Timeline of Life on Earth continued

3. Inferring Relationships How do mass extinctions appear to be related to the appearance of new major groups of organisms?

_____

_____

4. Justifying Conclusions Cyanobacteria are thought to be responsible for adding oxygen to Earth's atmosphere. Use your timeline to justify this conclusion.

_____

_____

_____

5. Calculating Determine the amount of time, as a percentage of the time that life has existed on Earth, that humans (Homo sapiens) have existed.

6. Further Inquiry Write a new question about the history of life on Earth that could be explored in another investigation.

_____

_____

_____

Name _____ Class _____ Date _____

Quick Lab

OBSERVATION

Analyzing Adaptations: Living on Land

The move from a watery environment to land was a giant step in evolution. In this lab, you will study some of the adaptations that allowed organisms to make the move to land.

OBJECTIVES

Observe various structural features of aquatic and terrestrial specimens.

Identify those structural features that are adaptations for life on land.

MATERIALS

- various aquatic and terrestrial specimens or photographs of specimens

Procedure

1. Observe each pair of specimens displayed around the classroom. One member of each pair is identified as an aquatic organism, and the other is identified as a terrestrial organism. Record the specimens' names and your observations of the specimens' structural features in **Table 1**. Use another sheet of paper if necessary.

TABLE 1 STRUCTURAL FEATURES

Specimen pairs		Observation of specimen's structural features	Adaptations for life on land
Name	Habitat		
1.	Aquatic		
	Terrestrial		
2.	Aquatic		
	Terrestrial		
3.	Aquatic		
	Terrestrial		
4.	Aquatic		
	Terrestrial		

Name _____ Class _____ Date _____

Analyzing Adaptations: Living on Land *continued*

2. For each pair of specimens, identify the structural features that are adaptations for life on land. Record the adaptations in **Table 1.**

Analysis and Conclusions

1. **Summarizing Data** List the structural features of terrestrial organisms that make them adapted for life on land.

   _____
   _____
   _____

2. **Analyzing Data** Choose one of the land adaptations you listed in **Table 1** and explain how it allows an organism to survive on land.

   _____
   _____
   _____
   _____

3. **Drawing Conclusions** Based on your observations, what are the two main obstacles that organisms had to overcome in order to survive on land?

   _____
   _____
   _____

4. **Interpreting Information** Why were animals unable to live on land before plants were able to?

   _____
   _____
   _____

5. **Drawing Conclusions** What can you conclude about the number of surviving offspring produced by the individuals that are best adapted to an environment?

   _____
   _____
   _____

Name _____ Class _____ Date _____

Skills Practice Lab

MATH/GRAPHING

Determining the Age of Artifacts Using C-14

Have you ever found an arrowhead or piece of pottery when you were walking through a recently plowed field or looking at a building excavation? Human-made objects, or artifacts, are sometimes found in such areas. Artifacts often are reminders that people from other cultures inhabited Earth thousands of years ago. If you can determine how old an artifact is and you know what cultures existed during the time the artifact was made, you might be able to determine what culture the object came from. This information can help you determine how those people lived.

How can artifacts be dated? Several methods are used, depending on how old the object is and its composition. Most of these methods are based on radioactivity. Many radioisotopes are present in nature and undergo radioactive decay. As they decay, they undergo changes in their nuclei, and they become different elements. As time passes, the amount of the radioisotope in a sample decreases at a regular rate, known as the radioisotope's half-life. One half-life is the amount of time it takes for half of the radioisotope in a sample to decay.

If an artifact was made from materials that were once living, its age can often be determined by using the radioisotope carbon-14 (C-14). The half-life of this carbon isotope is 5730 years. How can C-14 be used to determine the ages of organic artifacts? Most of the carbon dioxide in the atmosphere is formed from C-12, which is not radioactive. However, a small percentage of atmospheric carbon dioxide contains C-14 instead of C-12. Along with the carbon dioxide molecules containing C-12, these radioactive molecules are taken into green plants to be used during photosynthesis. While a plant lives, the C-14 decays, but it is also replaced. Thus, the amount of C-14 in the plant remains constant. Animals eat plants or other animals that eat plants. In this way, the C-14 in plants is passed along to animals. After a plant or animal dies, no more C-14 enters it. The C-14 present in the dead organic material continues to decay, so its amount decreases over time.

In this lab, you will determine the ages of various organic artifacts by comparing the amounts of C-14 remaining in those artifacts with the amounts that were initially present in them.

OBJECTIVES

Compute the ages of several artifacts from information provided and the half-life of C-14.

Determine the cultural sources of the artifacts by using a timeline.

MATERIALS

- calculator
- pen or pencil
- sheet of paper

Copyright © by Holt, Rinehart and Winston. All rights reserved.

Name _____ Class _____ Date _____

Determining the Age of Artifacts Using C-14 *continued*

Procedure

1. Examine **Table 1**. The artifacts listed there could have been discovered at a site in the Ohio River valley, where many prehistoric and historic native American civilizations made their homes. The timeline on this page and the next shows the native American civilizations that lived in the Ohio River valley and when they lived there.

2. On a separate sheet of paper, calculate the nearest whole number of half-lives that have passed since each artifact was made. Use the following equation, in which the variable n is the number of half-lives.

$$\left(\frac{1}{2}\right)^n = \frac{\text{current amount of C-14 in the artifact}}{\text{initial amount of C-14 in the artifact}}$$

For example, suppose by making comparisons with similar living materials, scientists determine that an artifact originally contained 2.964 g of C-14. When the artifact is dated, it contains 0.7235 g of C-14.

$$\left(\frac{1}{2}\right)^n = \frac{0.7235 \text{ g}}{2.964 \text{ g}} = 0.2441$$

For this example, the quotient, 0.2441, is approximately equal to $\frac{1}{4}$. Because $\frac{1}{4} = \left(\frac{1}{2}\right)\left(\frac{1}{2}\right) = \left(\frac{1}{2}\right)^2$, approximately two half-lives have passed.

3. Record the number of half-lives in **Table 2**. If the number cannot be determined, write "cannot determine" across the blank spaces for that artifact.

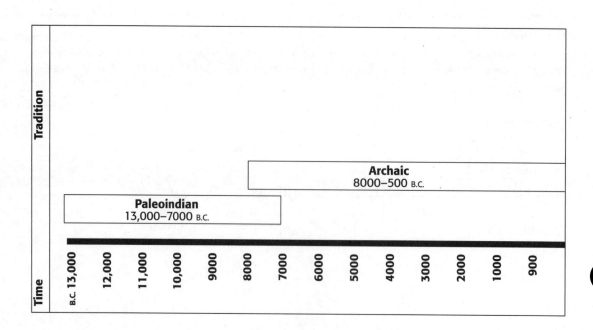

Determining the Age of Artifacts Using C-14 *continued*

4. Determine the age of each artifact, if it can be determined, from the number of half-lives that have passed since the artifact was made. To do this, multiply the number of half-lives by the half-life of C-14. For example, an artifact that has existed for three half-lives of C-14 is 3 × 5730 years, or 17,190 years old. Record your results in **Table 2**.

TABLE 1 INFORMATION ABOUT ARTIFACTS

Artifact	Estimated initial amount of C-14 present (g)	Current amount of C-14 present (g)
Cloth	1.436	0.353
Bone	2.343	none detected
Leather	1.322	0.08232
Wooden pole	0.05334	0.02592
Spearhead	0	0

TABLE 2 CONCLUSIONS ABOUT ARTIFACTS

Artifact	Number of half-lives	Age of artifact (yr)	Cultural source of artifact
Cloth			
Bone			
Leather			
Wooden pole			
Spearhead			

Tradition / Time

- Hopewell Culture: 100 B.C.–A.D. 400
- Adena Culture: 800 B.C.–A.D. 100
- Woodland: 800 B.C.–A.D. 1200
- Archaic: 8000–500 B.C.

Time scale: 800, 700, 600, 500, 400, 300, 200, 100 B.C., 0, A.D. 100, 200, 300, 400, 500

Determining the Age of Artifacts Using C-14 *continued*

Analysis

1. **Summarizing Data** What did you learn about the work of an archaeologist from doing this lab?

   _____
   _____
   _____

2. **Analyzing Data** After approximately nine half-lives have passed, too little C-14 remains in an object for the amount to be useful in dating the object. What is the minimum age of the bone listed in **Table 1**? Show your calculations.

   _____
   _____
   _____

3. **Analyzing Data** Compare the ages of the artifacts with the times when the cultures existed as shown on the timeline. From the calculated age of an artifact, decide which culture was the source of the artifact. Record your answers in **Table 2.**

Conclusions

1. **Drawing Conclusions** Why weren't the age and cultural source determined for the artifacts for which the number of half-lives could not be determined?

   _____
   _____

2. **Drawing Conclusions** Why was no C-14 present in the spearhead?

   _____
   _____

Extensions

1. **Research and Communications** Find out how C-14 dating has been used in determining the ages of relics, such as the ruins at Pompei. Prepare an oral or written report, including any effects the age has on the authenticity of the relic.

2. **Research and Communications** Contact a radiology department in a hospital to find out which radioisotopes are used in medical diagnoses and treatments. Find out the importance of using radioisotopes with short half-lives to minimize damage caused to cells in the human body by radiation.

Name _____ Class _____ Date _____

Quick Lab

DATASHEET FOR IN-TEXT LAB

Modeling Radioactive Decay

You can use some dried corn, a box, and a watch to make a model of radioactive decay that will show you how scientists measure the age of objects.

MATERIALS
- approximately 100 dry corn kernels per group
- cardboard box
- clock or watch with a second hand

Procedure

1. Assign one member of your team to keep time.
2. Place 100 dry corn kernels into a box.
3. Shake the box gently from side to side for 10 seconds.
4. Keep the box still and remove and count the kernels that "point" to the left side of the box. Record in the data table below the number of kernels you removed.

Data Table		
Total shake time (seconds)	Number of kernels removed	Number of kernels remaining
10		
20		
30		
40		
50		
60		

5. Repeat steps 4 and 5 until all kernels have been counted and removed.
6. Calculate the number of kernels remaining for each time interval.
7. Make a graph using your group's data. Plot "Total shake time (seconds)" on the x-axis. Plot "Number of kernels remaining" on the y-axis. Use a separate sheet of graph paper.

Modeling Radioactive Decay continued

Analysis

1. **Identify** what the removed kernels represent in each step.

   The kernels that are removed with each step represent the molecules that decayed.

2. **Calculate** the half-life of your sample, in seconds, that is represented in this activity.

   The half-life of the sample is the number of trials it took to remove 50 kernels multiplied by 10 sec/trial.

3. **Calculate** the age of your sample, in years, if each 10-second interval represents 5,700 years.

   Age of the sample = (number of trials) × (5,730 years/trial)

4. **Evaluate** the ability of this model to demonstrate radioactive decay.

   Answers will vary, but should be supported by the data.

Name _____ Class _____ Date _____

Quick Lab
Modeling Coacervates

DATASHEET FOR IN-TEXT LAB

By using simple chemistry, you will see that some properties of coacervates resemble the properties of cells.

MATERIALS
- safety goggles and lab apron
- graduated cylinder
- 1 percent gelatin solution
- 1 percent gum arabic solution
- test tube
- 0.1 M HCl
- pipet
- microscope slide and coverslip
- microscope

Procedure

1. **CAUTION: Hydrochloric acid is corrosive. Put on safety goggles, gloves, and a lab apron. Avoid contact with skin and eyes. Avoid breathing vapors. If any of this solution spills on you, immediately flush the area with water, and notify your teacher.**

2. Mix 5 mL of a 1 percent gelatin solution with 3 mL of a 1 percent gum arabic solution in a test tube.

3. Add 0.1 M HCl to the gelatin-gum arabic solution one drop at a time until the solution turns cloudy.

4. Prepare a wet mount of the cloudy solution, and examine it under a microscope at high power.

5. Prepare a drawing of the structures that you see.

TEACHER RESOURCE PAGE

Name _____ Class _____ Date _____

Modeling Coacervates *continued*

Analysis

1. **Describe** what happened to the solutions after the acid was added.

   The solution became cloudy.

2. **Compare** the appearance of coacervates with that of cells.

   Coacervates are spherical and membrane-bound but have no internal structures.

3. **Predict** what would happen to the coacervates if a base were added to the solution.

   Answers will vary but may include that coacervates might break up.

4. **Critical Thinking**
   Evaluating Hypotheses Based on the evidence you obtained, defend the hypothesis that coacervates could have been the basis of life on Earth

   Answers will vary. Students might state that coacervates are discrete structures that resemble cells and that arise in certain chemical environments.

TEACHER RESOURCE PAGE

Name _____ Class _____ Date _____

| Data Lab | DATASHEET FOR IN-TEXT LAB |

Analyzing Signs of Endosymbiosis

Background

You may recall that mitochondria have their own DNA and produce their own proteins. The data below were collected by scientists studying the proteins produced by mitochondrial DNA. The scientists found that the three-nucleotide sequences (codons) in the nucleus of an organism's cells can code for different amino acids than those coded for in the cell's mitochondria. Examine the data below, and answer the questions that follow.

Amino Acids Made in the Nucleus and Mitochondria			
	Amino acids or other instructions coded for in the nucleus	Amino acids or other instructions coded for in mitochondria	
Codon	Plants and mammals	Plants	Mammals
UGA	Stop	Stop	Tryptophan
AGA	Arginine	Arginine	Stop
AUA	Isoleucine	Isoleucine	Methionine
AUU	Isoleucine	Isoleucine	Methionine
CUA	Leucine	Leucine	Leucine

Analysis

1. **Defend** the theory of endosymbiosis using these data.

   The fact that in mammals the genetic code of nuclear DNA differs from that

   of mitochondrial DNA suggests that mitochondria have evolved separately

   from the animal cells that contain them.

2. **Infer** what these data indicate about the evolution of plant cells.

   We can infer that plant mitochondria and plant cells together have been

   subjected to selective pressures different from those that acted on

   mammalian cells and on mammalian mitochondria.

3. **Describe** how these data can be used to support the idea that more than one type of cell evolved early in the history of life.

   We can infer that the prokaryotic ancestors of mammalian mitochondria

   probably diverged very early from the prokaryotic ancestors of plant

   mitochondria because the genetic codes are quite different in the two types.

Copyright © by Holt, Rinehart and Winston. All rights reserved.

Holt Biology   History of Life on Earth

TEACHER RESOURCE PAGE

Name _____ Class _____ Date _____

Exploration Lab

DATASHEET FOR IN-TEXT LAB

Making a Timeline of Life on Earth

SKILLS
- Observing
- Inferring relationships
- Organizing data

OBJECTIVES
- **Compare** and **contrast** the distinguishing characteristics of representative organisms of the six kingdoms.
- **Organize** the appearance of life on Earth in a timeline.

MATERIALS
- adding-machine tape (5 m roll)
- meterstick
- colored pens or pencils
- photographs or drawings of organisms from ancient Earth to present day

Before You Begin

About 4.5 billion years ago, Earth was a ball of molten rock. As the surface cooled, a rocky crust formed and water vapor in the atmosphere condensed to form rain. By 3.9 billion years ago, oceans covered much of Earth's surface. Rocks formed in these oceans contain **fossils** of bacterial cells that lived about 3.5 billion years ago. The **fossil record** shows a progression of life-forms and contains evidence of many changes in Earth's surface and atmosphere.

In this lab, you will make a timeline showing the major events in Earth's history and in the history of life on Earth, such as the evolution of new groups of organisms and the mass extinctions. This timeline can be used to study how living things have changed over time.

1. Write a definition for each boldface term in the paragraphs above. Use a separate sheet of paper. **Answers appear in the TE for this lab.**

2. Record your data in the data table provided.

3. Based on the objectives for this lab, write a question you would like to explore about the history of life on Earth.

   Answers will vary. For example: How did life on Earth change after oxygen

   entered the atmosphere?

Copyright © by Holt, Rinehart and Winston. All rights reserved.
Holt Biology   History of Life on Earth

Name _____ Class _____ Date _____

Making a Timeline of Life on Earth *continued*

Procedure

PART A: MAKING A TIMELINE

1. Make a mark every 20 cm along a 5 m length of adding-machine tape. Label one end of the tape "5 billion years ago" and the other end "Today." Write "20 cm = 200 million years" near the beginning of your timeline.

2. Locate and label a point representing the origin of Earth on your timeline. Use your textbook as a reference. See the timeline at the bottom of Section 2 and Section 3 of this chapter. Also locate and label the 11 periods of the geologic time scale beginning with the Cambrian period. **Answers appear in the TE for this lab.**

3. Using your textbook as a reference, mark the following events on your timeline: the first cyanobacteria appear; oxygen enters the atmosphere; the five mass extinctions; the first eukaryotes appear; the first multicellular organisms appear; the first vertebrates appear; the first plants, fungi, and land animals appear; the first dinosaurs and mammals appear; the first flowering plants appear; the first humans appear. **Answers appear in the TE for this lab.**

4. Look at the photographs of organisms provided by your teacher. Identify the major characteristics of each organism. Record your observations in the data table below.

Data Table		
Organism	Kingdom	Characteristics/adaptation for life on Earth

Holt Biology — History of Life on Earth

TEACHER RESOURCE PAGE

Name _____ Class _____ Date _____

Making a Timeline of Life on Earth *continued*

5. Lay out your timeline on the floor in your classroom. Place photographs (or drawings) of the organisms you examined on your timeline to show when they appeared on Earth. **Answers appear in the TE for this lab.**

6. Fold the timeline at the mark representing 4.8 billion years ago. This leaves 24 segments, each representing 200 million years on your timeline. Now you can think of each segment as 1 hour in a 24-hour day.

7. When you are finished, walk slowly along your timeline. Note the sequence of events in the history of life on Earth and the relative amount of time between each event.

PART B: CLEANUP AND DISPOSAL

8. Dispose of paper scraps in the designated waste container.

9. Clean up your work area and all lab equipment. Return lab equipment to its proper place.

ANALYZE AND CONCLUDE

1. **Analyzing Information** Think of each segment of your timeline as 1 hour in a 24-hour day as you answer each of the following questions.

   a. How long has life existed on Earth?

   Life on Earth has existed for at least 17.5 hours.

   b. For what part of the day did only unicellular life-forms exist?

   Only unicellular life-forms existed for about 14 hours.

   c. At what time of day did the first plants appear on Earth?

   The first plants appeared at about 9:30 P.M.

   d. At what time of day did mammals appear on Earth?

   Mammals appeared at about 11:00 P.M.

2. **Summarizing Information** Identify the major developments in life-forms that have occurred over the last 3.5 billion years.

   Answers will vary. Life has changed from all prokaryotic, unicellular forms to complex multicellular forms composed of eukaryotic cells. Many life-forms that evolved have become extinct.

Copyright © by Holt, Rinehart and Winston. All rights reserved.
Holt Biology — History of Life on Earth

Name _____ Class _____ Date _____

Making a Timeline of Life on Earth *continued*

3. Inferring Relationships How do mass extinctions appear to be related to the appearance of new major groups of organisms?

Mass extinctions appear to be followed closely by the evolution of many new types of organisms.

4. Justifying Conclusions Cyanobacteria are thought to be responsible for adding oxygen to Earth's atmosphere. Use your timeline to justify this conclusion.

Cyanobacteria appeared before the accumulation of oxygen in the atmosphere, and cyanobacteria produce oxygen. Therefore, it is reasonable to conclude that they are responsible for adding oxygen to Earth's atmosphere.

5. Calculating Determine the amount of time, as a percentage of the time that life has existed on Earth, that humans (Homo sapiens) have existed.

Humans have existed for 0.014 percent of the time that life has existed on Earth.

6. Further Inquiry Write a new question about the history of life on Earth that could be explored in another investigation.

Answers will vary. Sample answer: How have Earth's climates changed since Precambrian times?

TEACHER RESOURCE PAGE

Quick Lab

OBSERVATION

Analyzing Adaptations: Living on Land

Teacher Notes

TIME REQUIRED 20 minutes

SKILLS ACQUIRED
Collecting data
Identifying patterns
Inferring
Organizing and analyzing data

RATINGS Easy ←—1—2—3—4—→ Hard
Teacher Prep–3
Student Setup–1
Concept Level–2
Cleanup–2

THE SCIENTIFIC METHOD

Make Observations Students observe the structural features of aquatic and terrestrial organisms.

Analyze the Results Analysis and Conclusions question 2 requires students to analyze their results.

Draw Conclusions Analysis and Conclusions questions 3 and 5 ask students to draw conclusions from their data.

MATERIALS

Materials for this lab can be purchased from WARD'S. See the *Master Materials List* for ordering instructions. Photographs may be used instead of actual specimens.

SAFETY CAUTIONS

- Discuss all safety symbols with students.
- Instruct students not to taste or eat any of the specimens.
- Remind students to handle animals carefully and with respect.

DISPOSAL

- Locally cultivated native plants may be introduced by replanting. Some states (such as California) have strict rules regarding the procurement or introduction of nonindigenous plants because of the possible presence of root-damaging nematodes. Usually, plants shipped to these states must pass inspection. Certain aquatic plants (*Elodea*, among others) should not be released or introduced into local habitats.

Analyzing Adaptations: Living on Land continued

- Permits are required to possess or release certain insects (such as cockroaches and termites). Check with your local office of the Animal and Plant Health Inspection Service (APHIS), the U.S. Department of Agriculture, or contact WARD'S.

- Nonindigenous animals should not be released into native habitats. In many cases, these organisms may not survive climatic conditions or may interfere with native fauna and flora. Contact your local APHIS office or WARD'S for specific information about whether a particular organism is considered nonindigenous to your area. In some cases, you may be able to find a home for an animal at a local pet shop.

- Eggs should be placed in sealed plastic bags and disposed of in a trash can.

TIPS AND TRICKS
Preparation
This lab works best in groups of two but can be done individually.

Discuss differences between terrestrial and aquatic environments.

Place pairs of specimens around the classroom for inspection. Each pair should consist of one aquatic specimen and one terrestrial specimen showing an adaptation for life on land. Include the name of each organism and any additional directions needed to draw students' attention to certain structural features.

A wide variety of specimens is available from WARD'S. Pairs of specimens might include any or all of the following:

- a chicken's egg and a frog's egg, to compare shelled with unshelled eggs
- an insect and a flatworm, jellyfish, or sponge, to highlight the cuticle or the exoskeleton of the insect
- a spore and a seed, to point out the seed coat and food supply of seeds
- an organism with lungs and one that uses diffusion or gills to obtain oxygen
- a bryophyte and a plant with an extensive root system

MISCONCEPTION ALERT

As you discuss adaptations, ask students if they have heard the phrase *survival of the fittest*, and if so, what they think it means. Some students might think it means that the bigger, stronger individuals always survive. Explain that *fit* means that the organisms are fit, or adapted, for survival in their environment, which may mean they are smaller, hide better, have thicker skin, or lay hard-shelled eggs. Being fit does not always mean being bigger or stronger.

TEACHER RESOURCE PAGE

Name _____ Class _____ Date _____

Quick Lab OBSERVATION

Analyzing Adaptations: Living on Land

The move from a watery environment to land was a giant step in evolution. In this lab, you will study some of the adaptations that allowed organisms to make the move to land.

OBJECTIVES

Observe various structural features of aquatic and terrestrial specimens.

Identify those structural features that are adaptations for life on land.

MATERIALS

- various aquatic and terrestrial specimens or photographs of specimens

Procedure

1. Observe each pair of specimens displayed around the classroom. One member of each pair is identified as an aquatic organism, and the other is identified as a terrestrial organism. Record the specimens' names and your observations of the specimens' structural features in **Table 1**. Use another sheet of paper if necessary.

TABLE 1 STRUCTURAL FEATURES

| Specimen pairs | | Observation of specimen's structural features | Adaptations for life on land |
Name	Habitat		
1.	Aquatic		
	Terrestrial		
2.	Aquatic		
	Terrestrial		
3.	Aquatic		
	Terrestrial		
4.	Aquatic		
	Terrestrial		

Copyright © by Holt, Rinehart and Winston. All rights reserved.

Holt Biology   History of Life on Earth

Analyzing Adaptations: Living on Land *continued*

2. For each pair of specimens, identify the structural features that are adaptations for life on land. Record the adaptations in **Table 1**.

Analysis and Conclusions

1. **Summarizing Data** List the structural features of terrestrial organisms that make them adapted for life on land.

   Answers may vary according to the specimens observed but might include

   roots, a waxy covering on plants, a waterproof exoskeleton, a seed with a

   seed coat and stored food, and lungs.

2. **Analyzing Data** Choose one of the land adaptations you listed in **Table 1** and explain how it allows an organism to survive on land.

   Accept any logical answer. For example, roots allow a plant to take in water

   from the soil; a waxy plant covering, a waterproof exoskeleton, and a seed

   coat all protect the organism from drying out; the stored food in a seed

   provides nutrients for the plant embryo until the seed germinates; and lungs

   allow an organism to take in oxygen from the air.

3. **Drawing Conclusions** Based on your observations, what are the two main obstacles that organisms had to overcome in order to survive on land?

   Desiccation and obtaining oxygen from the air are the two main obstacles

   that organisms had to overcome in order to survive on land.

4. **Interpreting Information** Why were animals unable to live on land before plants were able to?

   Unlike plants, animals are unable to make their own food. Animals feed on

   plants or on animals that eat plants.

5. **Drawing Conclusions** What can you conclude about the number of surviving offspring produced by the individuals that are best adapted to an environment?

   In most populations, the best-adapted individuals will produce the greatest

   number of surviving offspring.

TEACHER RESOURCE PAGE

Skills Practice Lab   MATH/GRAPHING
Determining the Age of Artifacts Using C-14

Teacher Notes

TIME REQUIRED One 45-minute period

SKILLS ACQUIRED
Collecting data
Inferring
Interpreting
Organizing and analyzing data

RATINGS   Easy ← 1   2   3   4 → Hard
Teacher Prep–1
Student Setup–1
Concept Level–3
Cleanup–1

THE SCIENTIFIC METHOD

Analyze the Results Analysis questions 2 and 3 require students to analyze their results.

Draw Conclusions Conclusions questions 1 and 2 ask students to draw conclusions.

Communicate the Results Analysis question 1 asks students to communicate their results.

MATERIALS

All calculations required in the lab can be done with any calculator. A scientific calculator is helpful if students wish to further investigate determining half-lives.

TECHNIQUES TO DEMONSTRATE

Demonstrate several examples of determining number of half-lives from initial and final amounts of radioisotopes.

TIPS AND TRICKS

This lab works best in groups of two but can be done individually.

Before the lab, familiarize students with the concept of half-life by using 100 coins as a model. This can be done in lab groups or as a classroom demonstration. The 100 coins represent the initial amount of a radioisotope present in an artifact. Place the coins in a box, and shake them well. Let the coins that have tails facing up represent the number of radioisotopes that have decayed. Count those coins, record the number, and remove them from the box. Repeat the procedure until all the coins have been removed. Students should determine that, if each trial represents a half-life, approximately half the radioisotopes decay from one half-life to the next.

Copyright © by Holt, Rinehart and Winston. All rights reserved.
Holt Biology   History of Life on Earth

Name _____ Class _____ Date _____

Skills Practice Lab   MATH/GRAPHING
Determining the Age of Artifacts Using C-14

Have you ever found an arrowhead or piece of pottery when you were walking through a recently plowed field or looking at a building excavation? Human-made objects, or artifacts, are sometimes found in such areas. Artifacts often are reminders that people from other cultures inhabited Earth thousands of years ago. If you can determine how old an artifact is and you know what cultures existed during the time the artifact was made, you might be able to determine what culture the object came from. This information can help you determine how those people lived.

How can artifacts be dated? Several methods are used, depending on how old the object is and its composition. Most of these methods are based on radioactivity. Many radioisotopes are present in nature and undergo radioactive decay. As they decay, they undergo changes in their nuclei, and they become different elements. As time passes, the amount of the radioisotope in a sample decreases at a regular rate, known as the radioisotope's half-life. One half-life is the amount of time it takes for half of the radioisotope in a sample to decay.

If an artifact was made from materials that were once living, its age can often be determined by using the radioisotope carbon-14 (C-14). The half-life of this carbon isotope is 5730 years. How can C-14 be used to determine the ages of organic artifacts? Most of the carbon dioxide in the atmosphere is formed from C-12, which is not radioactive. However, a small percentage of atmospheric carbon dioxide contains C-14 instead of C-12. Along with the carbon dioxide molecules containing C-12, these radioactive molecules are taken into green plants to be used during photosynthesis. While a plant lives, the C-14 decays, but it is also replaced. Thus, the amount of C-14 in the plant remains constant. Animals eat plants or other animals that eat plants. In this way, the C-14 in plants is passed along to animals. After a plant or animal dies, no more C-14 enters it. The C-14 present in the dead organic material continues to decay, so its amount decreases over time.

In this lab, you will determine the ages of various organic artifacts by comparing the amounts of C-14 remaining in those artifacts with the amounts that were initially present in them.

OBJECTIVES

Compute the ages of several artifacts from information provided and the half-life of C-14.

Determine the cultural sources of the artifacts by using a timeline.

MATERIALS

- calculator
- pen or pencil
- sheet of paper

Determining the Age of Artifacts Using C-14 continued

Procedure

1. Examine **Table 1**. The artifacts listed there could have been discovered at a site in the Ohio River valley, where many prehistoric and historic native American civilizations made their homes. The timeline on this page and the next shows the native American civilizations that lived in the Ohio River valley and when they lived there.

2. On a separate sheet of paper, calculate the nearest whole number of half-lives that have passed since each artifact was made. Use the following equation, in which the variable n is the number of half-lives.

$$\left(\frac{1}{2}\right)^n = \frac{\text{current amount of C-14 in the artifact}}{\text{initial amount of C-14 in the artifact}}$$

For example, suppose by making comparisons with similar living materials, scientists determine that an artifact originally contained 2.964 g of C-14. When the artifact is dated, it contains 0.7235 g of C-14.

$$\left(\frac{1}{2}\right)^n = \frac{0.7235 \text{ g}}{2.964 \text{ g}} = 0.2441$$

For this example, the quotient, 0.2441, is approximately equal to $\frac{1}{4}$. Because $\frac{1}{4} = \left(\frac{1}{2}\right)\left(\frac{1}{2}\right) = \left(\frac{1}{2}\right)^2$, approximately two half-lives have passed.

3. Record the number of half-lives in **Table 2**. If the number cannot be determined, write "cannot determine" across the blank spaces for that artifact.

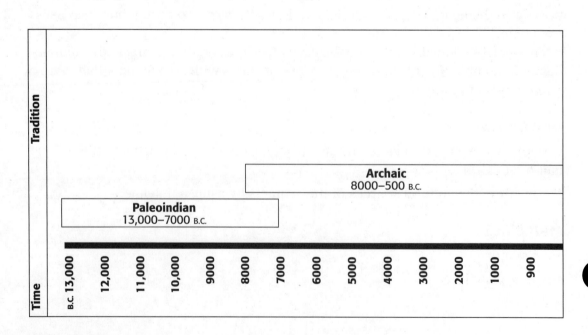

Determining the Age of Artifacts Using C-14 continued

4. Determine the age of each artifact, if it can be determined, from the number of half-lives that have passed since the artifact was made. To do this, multiply the number of half-lives by the half-life of C-14. For example, an artifact that has existed for three half-lives of C-14 is 3 × 5730 years, or 17,190 years old. Record your results in **Table 2**.

TABLE 1 INFORMATION ABOUT ARTIFACTS

Artifact	Estimated initial amount of C-14 present (g)	Current amount of C-14 present (g)
Cloth	1.436	0.353
Bone	2.343	none detected
Leather	1.322	0.08232
Wooden pole	0.05334	0.02592
Spearhead	0	0

TABLE 2 CONCLUSIONS ABOUT ARTIFACTS

Artifact	Number of half-lives	Age of artifact (yr)	Cultural source of artifact
Cloth	2	11,460	Paleoindian
Bone	cannot determine	cannot determine	cannot determine
Leather	4	22,920	cannot determine
Wooden pole	1	5730	Archaic
Spearhead	cannot determine	cannot determine	cannot determine

Hopewell Culture 100 B.C.–A.D. 400

Adena Culture 800 B.C.–A.D. 100

Woodland 800 B.C.–A.D. 1200

Archaic 8000–500 B.C.

Tradition / Time: 800, 700, 600, 500, 400, 300, 200, 100 B.C., 0, A.D. 100, 200, 300, 400, 500

Holt Biology — History of Life on Earth

TEACHER RESOURCE PAGE

Name _____ Class _____ Date _____

Determining the Age of Artifacts Using C-14 *continued*

Analysis

1. **Summarizing Data** What did you learn about the work of an archaeologist from doing this lab?

   Answers will vary. Students may conclude that information about ancient cultures has to be pieced together from available data and technological methods.

2. **Analyzing Data** After approximately nine half-lives have passed, too little C-14 remains in an object for the amount to be useful in dating the object. What is the minimum age of the bone listed in **Table 1**? Show your calculations.

   The minimum age of the bone would be approximately 50,000 years;

   9 × 5730 years = 51,570 years.

3. **Analyzing Data** Compare the ages of the artifacts with the times when the cultures existed as shown on the timeline. From the calculated age of an artifact, decide which culture was the source of the artifact. Record your answers in **Table 2**.

Conclusions

1. **Drawing Conclusions** Why weren't the age and cultural source determined for the artifacts for which the number of half-lives could not be determined?

   If the number of half-lives for an artifact could not be determined, the age of the artifact and its source also could not be determined with certainty.

2. **Drawing Conclusions** Why was no C-14 present in the spearhead?

   The spearhead contains no materials that were once living. Thus, it does not contain any C-14 from the atmosphere or from plants.

Extensions

1. **Research and Communications** Find out how C-14 dating has been used in determining the ages of relics, such as the ruins at Pompei. Prepare an oral or written report, including any effects the age has on the authenticity of the relic.

2. **Research and Communications** Contact a radiology department in a hospital to find out which radioisotopes are used in medical diagnoses and treatments. Find out the importance of using radioisotopes with short half-lives to minimize damage caused to cells in the human body by radiation.

Answer Key

Directed Reading

SECTION: HOW DID LIFE BEGIN?
1. radiometric dating
2. radioisotopes
3. half-life
4. The primordial soup model is the idea that the early Earth's oceans contained a soup of organic molecules that formed spontaneously through chemical reactions activated by solar radiation, volcanic eruptions, and lightning.
5. Recent discoveries have shown that the mixture of gasses used in Miller's experiment could not have existed on the early Earth. Then, ozone did not exist in the atmosphere, and ultraviolet radiation would have destroyed any ammonia and methane present. Without these gases, key biological molecules are not made.
6. The bubble model states that the processes that formed the chemicals needed for life took place within bubbles created by undersea volcanoes.
7. b
8. d
9. d
10. a
11. c
12. c

SECTION: THE EVOLUTION OF CELLULAR LIFE
1. The oldest fossils are remains of microscopic prokaryotes (bacteria). Cyanobacteria, which carried out photosynthesis and produced much of the oxygen in our atmosphere, were among the first prokaryotes to appear.
2. Eubacteria have peptidoglycan in their cell walls and have the same type of lipids in their cell membranes as eukaryotes have. Archaebacteria lack peptidoglycan in their cell walls and are thought to be closely related to the first bacteria that existed on Earth.
3. archaebacteria
4. eubacteria
5. The theory of endosymbiosis proposes that mitochondria and chloroplasts are the descendants of symbiotic aerobic eubacteria.
6. Chloroplasts are the organelles in plant cells where photosynthesis takes place.
7. d
8. c
9. c
10. b
11. a
12. Mass extinctions decreased competition for resources and therefore provided surviving species with a broad array of evolutionary opportunities.
13. because human activity is destroying Earth's ecosystems, such as rain forests

SECTION: LIFE INVADED THE LAND
1. 3
2. 1
3. 2
4. 5
5. 4
6. nutrients (food)
7. minerals
8. nutrients
9. minerals
10. mycorrhizae
11. mutualism
12. because animals need plants for food
13. Arthropods have a hard outer skeleton and jointed limbs.
14. lungs
15. amphibians
16. reptiles
17. continental drift
18. 2
19. 4
20. 1
21. 3

Active Reading

SECTION: HOW DID LIFE BEGIN?
1. ammonia, methane, and others
2. ultraviolet radiation and lightning

3. complex organic molecules
4. c

SECTION: COMPLEX ORGANISMS DEVELOP
1. Both are types of prokaryotes.
2. *Sulfolobus* is a living group of archaebacteria.
3. The cell walls of eubacteria contain peptidoglycan, and their cell membranes contain the same type of lipids found in eukaryotes.
4. The cell walls of archaebacteria lack peptidoglycan, and their cell membranes contain unique lipids.
5. The first bacteria to exist on Earth were closely related to archaebacteria.
6. The first eukaryotic cells likely evolved from archaebacteria.
7. a

SECTION: LIFE INVADED THE LAND
1. frogs, toads, and salamanders
2. several structural changes in their bodies
3. from the bones of fish fins
4. b

Vocabulary Review
1. A radioisotope is an unstable element that breaks up and gives off energy in the form of radiation; the half-life is the time it takes for half an amount of the element to change.
2. Fossils are the preserved or mineralized remains or imprint of an organism that lived long ago. Cyanobacteria are among the first bacteria to appear in the fossil record.
3. Eubacteria are prokaryotes whose cell walls contain a chemical called peptidoglycan; archaebacteria are prokaryotes that lack peptidoglycan in their cell walls and have unique lipids in their cell membrane.
4. Mutualism is a partnership in which both organisms benefit. An example is a mycorrhiza, which is the partnership between fungi and the roots of plants. The fungi provide minerals to the plant, and the plant provides nutrients to the fungi.
5. d
6. f
7. g
8. h
9. a
10. c
11. b
12. e

Science Skills
INTERPRETING TIMELINES
1. d
2. c
3. a
4. b
5. Humans are eliminating many species by destroying tropical rain forests.
6. Conditions on Earth have changed several times since Earth was formed. With each change, some organisms were better able to survive the new conditions than other organisms were. Some changes were brought about by geological and climate changes. These changes caused mass extinctions. The surviving organisms had less competition for resources.
7. Many scientists think a sixth mass extinction is already underway. Many species have recently become extinct because of human destruction of the tropical rain forests.
8. The 50 million years following the emergence of reptiles was a period of widespread drought. Reptiles, which are better adapted for dry conditions, had an apparent advantage over amphibians. Gradually, reptiles became the dominant animal group on Earth.
9. Cyanobacteria added oxygen to Earth's atmosphere, which made possible the development of a protective layer of ozone.

TEACHER RESOURCE PAGE

Concept Mapping

1. prokaryotes
2. fossils
3. 2.5 billion years ago
4. cyanobacteria
5. ozone
6. life on land
7. eubacteria
8. mitochondria or chloroplasts
9. chloroplasts or mitochondria
10. endosymbiosis

Critical Thinking

1. g
2. f
3. e
4. c
5. b
6. d
7. a
8. d
9. e
10. c
11. b
12. a
13. f
14. k, h
15. e, b
16. i, f
17. c, a
18. j, l
19. g
20. d
21. b
22. a
23. b
24. d

Test Prep Pretest

1. d
2. b
3. d
4. d
5. b
6. a
7. d
8. b
9. d
10. c
11. b
12. c
13. d
14. a
15. g
16. h
17. e
18. f
19. underwater bubbles
20. microspheres
21. animals
22. mutualism
23. arthropods
24. It allowed "division of labor" among cells and cell specialization, which led to organism complexity.
25. Amphibians, specifically the ancestors of today's frogs, toads, and salamanders, were the first vertebrates to live on land. Their lungs enabled them to absorb oxygen from the air. A strong, flexible internal skeleton made walking possible and allowed vertebrates to grow larger than insects.

Quiz

SECTION: HOW DID LIFE BEGIN?

1. b
2. a
3. c
4. d
5. f
6. b
7. d
8. e
9. a
10. c

SECTION: THE EVOLUTION OF CELLULAR LIFE

1. a
2. b
3. d
4. b
5. c
6. c
7. e
8. a
9. b
10. d

SECTION: LIFE INVADED THE LAND

1. c
2. a
3. d
4. c
5. b
6. e
7. c
8. b
9. a
10. d

Chapter Test (General)

1. d
2. e
3. a
4. c
5. b
6. d
7. d
8. b
9. b
10. d
11. c
12. a
13. b
14. c
15. c
16. d
17. b
18. c
19. d
20. b

Chapter Test (Advanced)

1. b
2. c
3. c
4. d
5. d
6. c
7. b
8. c
9. c
10. b
11. a
12. d
13. b
14. d
15. a
16. j
17. d
18. g
19. i
20. f
21. e
22. h
23. c
24. b
25. a

TEACHER RESOURCE PAGE

Lesson Plan

Section: How Did Life Begin?

Pacing

Regular Schedule: with lab(s): 4 days   without lab(s): 3 days

Block Schedule: with lab(s): 2 days   without lab(s): 1 1/2 days

Objectives

1. Summarize how radioisotopes can be used in determining Earth's age.
2. Compare two models that describe how the chemicals of life originated.
3. Describe how cellular organization might have begun.
4. Recognize the importance of the development of heredity to the development of life.

National Science Education Standards Covered

UNIFYING CONCEPTS AND PROCESSES

UCP1: Systems, order, and organization

UCP2: Evidence, models, and explanation

UCP3: Change, constancy, and measurement

UCP4: Evolution and equilibrium

UCP5: Form and function

SCIENCE AS INQUIRY

SI1: Abilities necessary to do scientific inquiry

SI2: Understandings about scientific inquiry

SCIENCE AND TECHNOLOGY

ST2: Understandings about science and technology

SCIENCE IN PERSONAL AND SOCIAL PERSPECTIVES

SPSP6: Science and technology in local, national, and global challenges

HISTORY AND NATURE OF SCIENCE

HNS1: Science as a human endeavor

HNS3: Historical perspectives

TEACHER RESOURCE PAGE

Lesson Plan *continued*

LIFE SCIENCE: THE CELL

LSCell6: Cells can differentiate, and complex multicellular organisms are formed as a highly organized arrangement of differentiated cells.

LIFE SCIENCE: BIOLOGICAL EVOLUTION

LSEvol1: Species evolve over time.

LSEvol2: The great diversity of organisms is the result of more than 3.5 billion years of evolution.

LSEvol3: Natural selection and its evolutionary consequences provide a scientific explanation for the fossil record of ancient life forms, as well as for the striking molecular similarities observed among the diverse species of living organisms.

LSEvol4: The millions of different species of plants, animals, and microorganisms that live on earth today are related by descent from common ancestors.

LIFE SCIENCE: MATTER, ENERGY, AND ORGANIZATION IN LIVING SYSTEMS

LSMat2: The energy for life primarily derives from the sun.

PHYSICAL SCIENCE

PS3: Chemical reactions

EARTH AND SPACE SCIENCE

ESS1: Energy in the Earth system

ESS3: Origin and evolution of the Earth system

ESS4: Origin and evolution of the universe

KEY
SE = Student Edition   **TE** = Teacher Edition
CRF = Chapter Resource File

Block 1

CHAPTER OPENER *(45 minutes)*

- **Quick Review,** SE. Students answer questions covered in previous sections of the textbook as preparation for the chapter content. (**GENERAL**)
- **Reading Activity,** SE. Students write down all the key words in the section and define them. (**GENERAL**)
- **Using the Figure,** TE. Students answer questions about the chapter opener photograph. (**GENERAL**)

TEACHER RESOURCE PAGE

Lesson Plan continued

- **Opening Activity**, TE. Students try to visualize the quantity one billion in terms of $1,000 bills. (**GENERAL**)

Block 2

FOCUS *(5 minutes)*

- **Bellringer Transparency.** Use this transparency as students enter the classroom and find their seats. (**GENERAL**)

MOTIVATE *(5 minutes)*

- **Using the Figure**, TE. Students explain why the rate of radioactive decay (shown in Figure 1) is not a straight line. (**GENERAL**)

TEACH *(35 minutes)*

- **Teaching Transparency, Section Outline.** Use this transparency to give students a framework for the information in this section. (**GENERAL**)
- **Teaching Transparency, Radioactive Decay.** Use this transparency to show students the rate of decay for the radioisotope potassium-40. Review the terms radioisotopes and half-life. Have students use this graph to estimate the age of rocks that contain potassium-40. (**GENERAL**)
- **Inclusion Strategies**, TE. Students draw and narrate illustrations of their interpretation of Earth's formation 4.5 billion years ago. (**BASIC**)
- **Quick Lab,** Modeling Radioactive Decay, SE. Using corn to model radioactive decay, students see how scientists measure the age of objects. (**GENERAL**)
- **Datasheets for In-Text Labs,** Modeling Radioactive Decay, CRF.

HOMEWORK

- **Directed Reading Worksheet, How Did Life Begin?, CRF.** Students complete the exercises in this worksheet to help them understand the material as they read the section. (**BASIC**)
- **Active Reading Worksheet, How Did Life Begin?, CRF.** Students read a passage related to the section topic and answer questions. (**GENERAL**)

Block 3

TEACH *(40 minutes)*

- **Teaching Transparency, Miller-Urey Experiment.** Use this transparency to walk students through the device used in the Miller-Urey experiment. (**GENERAL**)
- **Integrating Physics and Chemistry**, TE. Students identify the number of valence electrons in the components of the "primordial soup" and relate this information to bonding pattern. (**BASIC**)

TEACHER RESOURCE PAGE

Lesson Plan *continued*

- **Teaching Transparency, Lerman's Bubble Model.** Use this transparency to walk students through the main points in Lerman's hypothesis, which explains the origin of life's basic chemicals. **(GENERAL)**
- **Quick Lab,** Modeling Coacervates, SE. Students create coacervates to see how they may have led to the first cells. **(GENERAL)**
- **Datasheets for In-Text Labs,** Modeling Coacervates, CRF.

CLOSE (5 minutes)

- **Quiz,** TE. Students answer questions that review the section material. **(GENERAL)**

HOMEWORK

- **Reteaching,** TE. Students draw and label the parts of the Miller-Urey apparatus. **(BASIC)**
- **Section Review,** SE. Assign questions 1–5 for review, homework, or quiz. **(GENERAL)**
- **Alternative Assessment,** TE. Students write an answer to the following question: "Were the first cells that evolved on Earth prokaryotes or eukaryotes?" Have the students justify their choice. **(GENERAL)**
- **Quiz, CRF.** This quiz consists of ten multiple choice and matching questions that review the section's main concepts. **(BASIC) Also in Spanish.**

Optional Block

LAB (45 minutes)

- **Skills Practice Lab, Determining the Age of Artifacts Using C-14, CRF.** Students determine the ages of various organic artifacts by comparing the amounts of carbon-14 remaining in those artifacts with the amounts that were initially present in them. **(GENERAL)**

Other Resource Options

- **Internet Connect.** Students can research Internet sources about Radioactive Decay with SciLinks Code HX4154.
- **go.hrw.com.** For worksheets, videos, and other teaching aids related to this chapter, visit the HRW Web site and type in the keyword HX4 LIF.
- **CNN Science in the News, Video Segment 8 Dino Egg Discovery.** This video segment is accompanied by a **Critical Thinking Worksheet**.
- **CNN Student News.** Find the latest news, lesson plans, and activities related to important scientific events at **cnnstudentnews.com**.

TEACHER RESOURCE PAGE
Lesson Plan

Section: The Evolution of Cellular Life

Pacing
Regular Schedule:　　with lab(s): N/A　　without lab(s): 2 days
Block Schedule:　　with lab(s): N/A　　without lab(s): 1 day

Objectives
1. Distinguish between the two groups of prokaryotes.
2. Describe the evolution of eukaryotes.
3. Recognize an evolutionary advance first seen in protists.
4. Summarize how mass extinctions have affected the evolution of life on Earth.

National Science Education Standards Covered
UNIFYING CONCEPTS AND PROCESSES
UCP1: Systems, order, and organization

UCP2: Evidence, models, and explanation

UCP3: Change, constancy, and measurement

UCP4: Evolution and equilibrium

UCP5: Form and function

SCIENCE AS INQUIRY
SI1: Abilities necessary to do scientific inquiry

SI2: Understandings about scientific inquiry

SCIENCE AND TECHNOLOGY
ST2: Understandings about science and technology

HISTORY AND NATURE OF SCIENCE
HNS1: Science as a human endeavor

HNS3: Historical perspectives

LIFE SCIENCE: THE CELL
LSCell6: Cells can differentiate, and complex multicellular organisms are formed as a highly organized arrangement of differentiated cells.

LIFE SCIENCE: BIOLOGICAL EVOLUTION
LSEvol1: Species evolve over time.

Copyright © by Holt, Rinehart and Winston. All rights reserved.

TEACHER RESOURCE PAGE
Lesson Plan continued

LSEvol2: The great diversity of organisms is the result of more than 3.5 billion years of evolution.

LIFE SCIENCE: MATTER, ENERGY, AND ORGANIZATION IN LIVING SYSTEMS

LSMat2: The energy for life primarily derives from the sun.

LIFE SCIENCE: BEHAVIOR OF ORGANISMS

LSBeh3: Like other aspects of an organism's biology, behaviors have evolved through natural selection.

PHYSICAL SCIENCE

PS3: Chemical reactions

PS6: Interactions of energy and matter

KEY
SE = Student Edition   TE = Teacher Edition
CRF = Chapter Resource File

Block 4

FOCUS *(5 minutes)*

- **Bellringer Transparency.** Use this transparency as students enter the classroom and find their seats. **(GENERAL)**

MOTIVATE *(10 minutes)*

- **Demonstration**, TE. Draw a large circle on the board to represent a clock. Place an arrow at corresponding times on the clock that note when certain life-forms first appeared. **(GENERAL)**

TEACH *(30 minutes)*

- **Teaching Transparency, Section Outline.** Use this transparency to give students a framework for the information in this section. **(GENERAL)**
- **Teaching Transparency, Evolution of Eukaryotes.** Use this transparency to discuss the evolution of eukaryotes. **(GENERAL)**
- **Data Lab,** Analyzing Signs of Endosymbiosis, SE. Students analyze a data table of amino acids produced in the nucleus and mitochondria. **(GENERAL)**

HOMEWORK

- **Directed Reading Worksheet, The Evolution of Cellular Life, CRF.** Students complete the exercises in this worksheet to help them understand the material as they read the section. **(BASIC)**

Copyright © by Holt, Rinehart and Winston. All rights reserved.

TEACHER RESOURCE PAGE

Lesson Plan *continued*

- **Active Reading Worksheet,** The Evolution of Cellular Life, CRF. Students read a passage related to the section topic and answer questions. (**GENERAL**)

Block 5

TEACH *(30 minutes)*

- **Group Activity,** Models of Endosymbionts and Bacteria, TE. Students create models or posters of chloroplasts, mitochondria, non-photosynthetic bacteria and photosynthetic bacteria. (**GENERAL**)
- **Demonstration,** TE. Use examples of fossils for identification or classification activities. (**GENERAL**)
- **Teaching Tip,** Endosymbiosis, TE. Review other examples of endosymbiosis, such as bacteria within the digestive tracts of termites. (**GENERAL**)

CLOSE *(15 minutes)*

- **Alternative Assessment,** TE. Students create their own quiz show using true and false questions dealing with the section material. (**GENERAL**)
- **Quiz,** TE. Students answer questions that review the section material. (**GENERAL**)

HOMEWORK

- **Reteaching,** TE. Students write a one-page summary of the section and make a timeline. (**BASIC**)
- **Activity,** Favorite Fossils, TE. Students write a short report on an extinct species of their choosing. (**ADVANCED**)
- **Section Review,** SE. Assign questions 1–6 for review, homework, or quiz. (**GENERAL**)
- **Quiz, CRF.** This quiz consists of ten multiple choice and matching questions that review the section's main concepts. (**BASIC**) **Also in Spanish.**

Other Resource Options

- **Activity,** Burgess Shale, TE. Students resarch the Burgess Shale and make simple sketches of organisms that would be considered unusual today. (**ADVANCED**)
- **Internet Connect.** Students can research Internet sources about Extinction with SciLinks Code HX4078.
- **Internet Connect.** Students can research Internet sources Endosymbiosis with SciLinks Code HX4068.
- **go.hrw.com.** For worksheets, videos, and other teaching aids related to this chapter, visit the HRW Web site and type in the keyword HX4 LIF.
- **CNN Student News.** Find the latest news, lesson plans, and activities related to important scientific events at **cnnstudentnews.com**.

TEACHER RESOURCE PAGE

Lesson Plan

Section: Life Invaded the Land

Pacing
Regular Schedule: with lab(s): 4 days without lab(s): 2 days
Block Schedule: with lab(s): 2 days without lab(s): 1 day

Objectives
1. Relate the development of ozone to the adaptation of life to the land.
2. Identify the first multicellular organisms to live on land.
3. Name the first animals to live on land.
4. List the first vertebrates to leave the oceans.

National Science Education Standards Covered

UNIFYING CONCEPTS AND PROCESSES

UCP3: Change, constancy, and measurement

UCP4: Evolution and equilibrium

SCIENCE IN PERSONAL AND SOCIAL PERSPECTIVES

SPSP5: Natural and human-induced hazards

HISTORY AND NATURE OF SCIENCE

HNS3: Historical perspectives

LIFE SCIENCE: THE CELL

LSCell6: Cells can differentiate, and complex multicellular organisms are formed as a highly organized arrangement of differentiated cells.

LIFE SCIENCE: BIOLOGICAL EVOLUTION

LSEvol1: Species evolve over time.

LSEvol2: The great diversity of organisms is the result of more than 3.5 billion years of evolution.

LSEvol3: Natural selection and its evolutionary consequences provide a scientific explanation for the fossil record of ancient life forms, as well as for the striking molecular similarities observed among the diverse species of living organisms.

LSEvol4: The millions of different species of plants, animals, and microorganisms that live on earth today are related by descent from common ancestors.

Copyright © by Holt, Rinehart and Winston. All rights reserved.

Holt Biology — History of Life on Earth

TEACHER RESOURCE PAGE

Lesson Plan *continued*

LIFE SCIENCE: BEHAVIOR OF ORGANISMS

LSBeh2: Organisms have behavioral responses to internal changes and to external stimuli.

KEY
SE = Student Edition TE = Teacher Edition
CRF = Chapter Resource File

Block 6

FOCUS *(5 minutes)*

- **Bellringer Transparency.** Use this transparency as students enter the classroom and find their seats. (**GENERAL**)

MOTIVATE *(10 minutes)*

- **Discussion/Question**, TE. Ask students if ozone is a desirable part of our atmosphere. Then ask if ozone is always a desirable part of our atmosphere. (**GENERAL**)

TEACH *(30 minutes)*

- **Teaching Transparency, Section Outline.** Use this transparency to give students a framework for the information in this section. (**GENERAL**)
- **Teaching Transparency, Ozone Shields the Earth.** Use this transparency to discuss how ozone formed as cyanobacteria added oxygen to the atmosphere. (**GENERAL**)
- **Inclusion Strategies**, TE. Students create their own resource documents of animal classification. (**BASIC**)
- **Demonstration**, TE. Students crack open an egg in a Petri dish and place an intact egg in another dish, crack open the intact egg two days later and weigh each egg separately, and explain the difference in the weights of the eggs. (**BASIC**)

HOMEWORK

- **Directed Reading Worksheet, Life Invaded the Land, CRF.** Students complete the exercises in this worksheet to help them understand the material as they read the section. (**BASIC**)
- **Active Reading Worksheet, Life Invaded the Land, CRF.** Students read a passage related to the section topic and answer questions. (**GENERAL**)

Block 7

TEACH *(30 minutes)*

- **Demonstration**, TE. Students use soft, gelatin-based candy worms, toothpicks, and tape to create a model of the internal and external skeletons of worms. (**GENERAL**)

Copyright © by Holt, Rinehart and Winston. All rights reserved.

TEACHER RESOURCE PAGE

Lesson Plan *continued*

- **Quick Lab, Analyzing Adaptations: Living on Land, CRF.** Students observe various structural features of aquatic and terrestrial specimens and identify the features that are adaptations for living on land. (**GENERAL**)

CLOSE *(15 minutes)*

- **Science Skills Worksheet, CRF.** Students interpret a timeline of the history of life on Earth. (**GENERAL**)
- **Quiz,** TE. Students answer questions that review the section material. (**GENERAL**)

HOMEWORK

- **Reteaching,** TE. Students complete a table that summarizes the evolution of vertebrates. (**BASIC**)
- **Alternative Assessment,** TE. Students make a model of a eukaryotic cell using different foods. (**GENERAL**)
- **Section Review,** SE. Assign questions 1–6 for review, homework, or quiz. (**GENERAL**)
- **Quiz, CRF.** This quiz consists of ten multiple choice and matching questions that review the section's main concepts. (**BASIC**) **Also in Spanish.**
- **Modified Worksheet, One-Stop Planner.** This worksheet has been specially modified to reach struggling students. (**BASIC**)
- **Critical Thinking Worksheet, CRF.** Students answer analogy-based questions that review the section's main concepts and vocabulary. (**ADVANCED**)

Optional Blocks

LAB *(90 minutes)*

- **Exploration Lab,** Making a Timeline of Life on Earth, SE. Students construct timelines showing geologic time and the major events in the evolution of life on Earth. (**GENERAL**)
- **Datasheets for In-Text Labs, Making a Timeline of Life on Earth, CRF.**

Other Resource Options

- **Supplemental Reading, The Dinosaur Heresies, One-Stop Planner.** Students read the book and answer questions. (**ADVANCED**)
- **Internet Connect.** Students can research Internet sources about Extinction with SciLinks Code HX4078.
- **go.hrw.com.** For worksheets, videos, and other teaching aids related to this chapter, visit the HRW Web site and type in the keyword HX4 LIF.
- **CNN Science in the News, Video Segment 8 Dino Egg Discovery.** This video segment is accompanied by a **Critical Thinking Worksheet**.
- **CNN Student News.** Find the latest news, lesson plans, and activities related to important scientific events at **cnnstudentnews.com**.

TEACHER RESOURCE PAGE

Lesson Plan

End-of-Chapter Review and Assessment

Pacing
Regular Schedule: 2 days
Block Schedule: 1 day

KEY
SE = Student Edition   **TE** = Teacher Edition
CRF = Chapter Resource File

Block 8
REVIEW *(45 minutes)*

- **Study Zone,** SE. Use the Study Zone to review the Key Concepts and Key Terms of the chapter and prepare students for the Performance Zone questions. (**GENERAL**)

- **Performance Zone,** SE. Assign questions to review the material for this chapter. Use the assignment guide to customize review for sections covered. (**GENERAL**)

- **Teaching Transparency, Concept Mapping.** Use this transparency to review the concept map for this chapter. (**GENERAL**)

Block 9
ASSESSMENT *(45 minutes)*

- **Chapter Test, History of Life on Earth, CRF.** This test contains 20 multiple choice and matching questions keyed to the chapter's objectives. (**GENERAL**) **Also in Spanish.**

- **Chapter Test, History of Life on Earth, CRF.** This test contains 25 questions of various formats, each keyed to the chapter's objectives. (**ADVANCED**)

- **Modified Chapter Test, One-Stop Planner.** This test has been specially modified to reach struggling students. (**BASIC**)

Other Resource Options

- **Vocabulary Review Worksheet, CRF.** Use this worksheet to review the chapter vocabulary. (**GENERAL**) **Also in Spanish.**

- **Test Prep Pretest, CRF.** Use this pretest to review the main content of the chapter. Each question is keyed to a section objective. (**GENERAL**) **Also in Spanish.**

- **Test Item Listing for ExamView® Test Generator, CRF.** Use the Test Item Listing to identify questions to use in a customized homework, quiz, or test.

- **ExamView® Test Generator, One-Stop Planner.** Create a customized homework, quiz, or test using the HRW Test Generator program.

TEST ITEM LISTING

History of Life on Earth

TRUE/FALSE

1. ____ Scientists are unable to calculate the age of Earth.
   Answer: False   Difficulty: I   Section: 1   Objective: 1

2. ____ Radiometric dating measures the age of an object by measuring the proportions of radioactive isotopes.
   Answer: True   Difficulty: I   Section: 1   Objective: 1

3. ____ The half-life of potassium-40 is 5,700 years.
   Answer: False   Difficulty: I   Section: 1   Objective: 1

4. ____ Radioisotopes are stable elements.
   Answer: False   Difficulty: I   Section: 1   Objective: 1

5. ____ On the early Earth, oxygen was found in the atmosphere millions of years before it was present in the oceans.
   Answer: False   Difficulty: I   Section: 1   Objective: 2

6. ____ The primordial soup model states that life developed when molecules of nonliving matter reacted chemically, forming simple organic molecules.
   Answer: True   Difficulty: I   Section: 1   Objective: 2

7. ____ Louis Lerman is credited with developing the bubble model for the origin of life's chemicals.
   Answer: True   Difficulty: I   Section: 1   Objective: 2

8. ____ The origin of cells is clearly understood.
   Answer: False   Difficulty: I   Section: 1   Objective: 3

9. ____ Scientists hypothesize that DNA may have acted as a catalyst in the formation of the first proteins.
   Answer: False   Difficulty: I   Section: 1   Objective: 3

10. ____ Microspheres are tiny vesicles formed by groups of short chains of amino acids.
    Answer: True   Difficulty: I   Section: 1   Objective: 3

11. ____ Scientists think that formation of fatty acids might have been the first step toward cellular organization.
    Answer: False   Difficulty: I   Section: 1   Objective: 3

12. ____ Double-stranded DNA evolved before RNA.
    Answer: False   Difficulty: I   Section: 1   Objective: 4

13. ____ Life began before the mechanism of heredity was developed.
    Answer: False   Difficulty: I   Section: 1   Objective: 4

14. ____ Eukaryotes are characterized by an internal membrane system and a nucleus containing DNA.
    Answer: True   Difficulty: I   Section: 1   Objective: 4

15. ____ The first living things to appear on Earth were prokaryotes.
    Answer: True   Difficulty: I   Section: 2   Objective: 1

16. ____ The photosynthetic cyanobacteria produced the oxygen in Earth's atmosphere.
    Answer: True   Difficulty: I   Section: 2   Objective: 1

17. ____ The two major groups of bacteria are eubacteria and cyanobacteria.
    Answer: False   Difficulty: I   Section: 2   Objective: 1

Copyright © by Holt, Rinehart and Winston. All rights reserved.

Holt Biology

TEST ITEM LISTING, continued

18. ____ Eukaryotes probably descended from bacteria.
    Answer: True   Difficulty: I   Section: 2   Objective: 2

19. ____ Both mitochondria and nuclei are believed to have their beginnings as prokaryotic parasites that invaded the pre-eukaryotic cell.
    Answer: False   Difficulty: I   Section: 2   Objective: 2

20. ____ The theory of endosymbiosis proposes that mitochondria and chloroplasts are the descendants of symbiotic aerobic eubacteria.
    Answer: True   Difficulty: I   Section: 2   Objective: 2

21. ____ Mitochondria and chloroplasts each contain DNA unrelated to the nuclear DNA of the cells in which they reside.
    Answer: True   Difficulty: I   Section: 2   Objective: 2

22. ____ Some protists are photosynthetic.
    Answer: True   Difficulty: I   Section: 2   Objective: 3

23. ____ All protists are single-celled.
    Answer: False   Difficulty: I   Section: 2   Objective: 3

24. ____ Multicellularity was not important to the evolution of life on Earth.
    Answer: False   Difficulty: I   Section: 2   Objective: 3

25. ____ Algae are members of the kingdom Plantae.
    Answer: False   Difficulty: I   Section: 2   Objective: 3

26. ____ Multicellular protists evolved independently many different times.
    Answer: True   Difficulty: I   Section: 2   Objective: 3

27. ____ Mass extinctions occur with regularity every 125 million years.
    Answer: False   Difficulty: I   Section: 2   Objective: 4

28. ____ The first four mass extinctions were likely caused by global geological and weather changes.
    Answer: True   Difficulty: I   Section: 2   Objective: 4

29. ____ Mass extinctions have had a significant impact on the course of evolution of life on Earth.
    Answer: True   Difficulty: I   Section: 2   Objective: 4

30. ____ Ozone was not present approximately 4 billion years ago.
    Answer: True   Difficulty: I   Section: 3   Objective: 1

31. ____ Life on land could not exist without ozone.
    Answer: True   Difficulty: I   Section: 3   Objective: 1

32. ____ Oxygen, O_2, reacts with carbon dioxide in the upper atmosphere to form ozone, O_3.
    Answer: False   Difficulty: I   Section: 3   Objective: 1

33. ____ Plants and algae developed a mutualistic relationship that allowed them to be the first organisms to live on land.
    Answer: False   Difficulty: I   Section: 3   Objective: 2

34. ____ Plants have been on land for more than 1 billion years.
    Answer: False   Difficulty: I   Section: 3   Objective: 2

35. ____ Plants can extract minerals from bare rock.
    Answer: False   Difficulty: I   Section: 3   Objective: 2

36. ____ The evolution of land plants had to take place before the evolution of land animals.
    Answer: True   Difficulty: I   Section: 3   Objective: 2

Holt Biology — History of Life on Earth

TEST ITEM LISTING, continued

37. ____ Flight enabled insects to search Earth's surface for food, mates, and nesting sites.
    Answer: True   Difficulty: I   Section: 3   Objective: 3

38. ____ The first animals to live on land were birds.
    Answer: False   Difficulty: I   Section: 3   Objective: 3

39. ____ Insects evolved from aquatic arthropods.
    Answer: True   Difficulty: I   Section: 3   Objective: 3

40. ____ All arthropods can fly.
    Answer: False   Difficulty: I   Section: 3   Objective: 3

41. ____ Arthropods evolved from insects.
    Answer: False   Difficulty: I   Section: 3   Objective: 3

42. ____ The distinguishing feature of vertebrates is that none of them have jaws.
    Answer: False   Difficulty: I   Section: 3   Objective: 4

43. ____ The first fish to evolve had small, powerful jaws.
    Answer: False   Difficulty: I   Section: 3   Objective: 4

44. ____ Amphibians can lay their eggs on dry land because the eggs are surrounded by a shell that prevents water loss.
    Answer: False   Difficulty: I   Section: 3   Objective: 4

45. ____ Reptiles thrive in dry climates.
    Answer: True   Difficulty: I   Section: 3   Objective: 4

46. ____ Amphibians were able to become successful on land because their skin is watertight.
    Answer: False   Difficulty: I   Section: 3   Objective: 4

MULTIPLE CHOICE

47. The age of Earth is estimated at about
    a. 200,000 years.
    b. 4.5 billion years.
    c. 2 million years.
    d. 2 billion years.
    Answer: B   Difficulty: I   Section: 1   Objective: 1

48. What percentage of potassium-40 remains after two half-lives?
    a. 100%
    b. 75%
    c. 50%
    d. 25%
    Answer: D   Difficulty: I   Section: 1   Objective: 1

49. How long is the half-life of potassium-40?
    a. 1.3 billion years
    b. 2.6 billion years
    c. 3.9 billion years
    d. 5.2 billion years
    Answer: A   Difficulty: I   Section: 1   Objective: 1

50. As the amount of potassium-40 decreases, the amount of stable isotope formed
    a. decreases.
    b. increases.
    c. increases then decreases.
    d. remains the same.
    Answer: B   Difficulty: I   Section: 1   Objective: 1

51. Energy used in the formation of the first organic molecules is thought to have come from
    a. water.
    b. the sun.
    c. air.
    d. fire.
    Answer: B   Difficulty: I   Section: 1   Objective: 2

TEST ITEM LISTING, continued

52. The model stating that organic molecules present in ancient seas led to the formation of life's building blocks was the
    a. radioactive decay model.
    b. bubble model.
    c. primordial soup model.
    d. None of the above

    Answer: C   Difficulty: I   Section: 1   Objective: 2

53. The Lerman bubble model proposes that
    a. ammonia and methane gases were trapped in underwater bubbles.
    b. gases reacted within bubbles, producing organic molecules.
    c. organic molecules were released into the air when the bubbles popped.
    d. All of the above

    Answer: D   Difficulty: I   Section: 1   Objective: 2

54. The early molecular catalyst that may have assisted in building the first proteins was
    a. RNA.
    b. an amino acid.
    c. DNA.
    d. a lipid.

    Answer: A   Difficulty: I   Section: 1   Objective: 3

55. The first step towards cellular organization may have come in the form of
    a. microsatellites.
    b. microspheres.
    c. micrometers.
    d. micromolecules.

    Answer: B   Difficulty: I   Section: 1   Objective: 3

56. Cyanobacteria are thought to be the ancestors of
    a. mitochondria.
    b. nuclei.
    c. ribosomes.
    d. chloroplasts.

    Answer: D   Difficulty: I   Section: 2   Objective: 1

57. Some archaebacteria
    a. cause most of the diseases found in people today.
    b. often cause decay.
    c. have ways of producing energy without using oxygen.
    d. are important in the production of cheese and other dairy products.

    Answer: C   Difficulty: I   Section: 2   Objective: 1

58. Biologists separate bacteria into two groups based upon
    a. the composition of their cell walls.
    b. the structure of some of their proteins.
    c. the chemical composition of their cell membranes.
    d. All of the above

    Answer: D   Difficulty: I   Section: 2   Objective: 1

59. Cyanobacteria changed the young Earth's atmosphere by producing
    a. carbon dioxide.
    b. ammonia.
    c. methane.
    d. oxygen.

    Answer: D   Difficulty: I   Section: 2   Objective: 1

60. Cyanobacteria changed Earth's atmosphere as they carried out the process of
    a. atmospheric bonding.
    b. nitrogen synthesis.
    c. photosynthesis.
    d. gradualism.

    Answer: C   Difficulty: I   Section: 2   Objective: 1

61. archaebacteria : eukaryotes ::
    a. eubacteria : cyanobacteria
    b. archaebacteria : eubacteria
    c. eubacteria : archaebacteria
    d. cyanobacteria : eubacteria

    Answer: D   Difficulty: II   Section: 2   Objective: 1

TEST ITEM LISTING, continued

62. Eukaryotes may have descended from
    a. eubacteria.
    b. archaebacteria.
    c. cyanobacteria.
    d. None of the above
    Answer: B   Difficulty: I   Section: 2   Objective: 2

63. Eukaryotes first appeared
    a. 1.5 million years ago.
    b. 150 million years ago.
    c. 1.5 billion years ago.
    d. 150 billion years ago.
    Answer: C   Difficulty: I   Section: 2   Objective: 2

64. Pre-eukaryotic cells did *not* contain
    a. mitochondria.
    b. cell membranes.
    c. DNA.
    d. RNA.
    Answer: C   Difficulty: I   Section: 2   Objective: 2

65. Chloroplasts are thought to be the result of an invasion of pre-eukaryotic cells by
    a. mitochondria.
    b. ribosomes.
    c. photosynthetic bacteria.
    d. protists.
    Answer: C   Difficulty: I   Section: 2   Objective: 2

66. Chloroplasts and mitochondria are both thought to have evolved through the process of
    a. photosynthesis.
    b. cellular respiration.
    c. chemical reactions.
    d. endosymbiosis.
    Answer: D   Difficulty: I   Section: 2   Objective: 2

67. Protists evolved from
    a. archaebacteria.
    b. protobacteria.
    c. eubacteria.
    d. pseudobacteria.
    Answer: A   Difficulty: I   Section: 2   Objective: 3

68. The first eukaryotic kingdom was the Kingdom
    a. Animalia.
    b. Plantae.
    c. Protista.
    d. Fungi.
    Answer: C   Difficulty: I   Section: 2   Objective: 3

69. The most diverse eukaryotic kingdom is
    a. Protista.
    b. Plantae.
    c. Animalia.
    d. Fungi.
    Answer: A   Difficulty: I   Section: 2   Objective: 3

70. The oldest known fossils of multicellular organisms were found in rocks that were
    a. 100 million years old.
    b. 700 million years old.
    c. 1.5 billion years old.
    d. 4 billion years old.
    Answer: B   Difficulty: I   Section: 2   Objective: 3

71. The kingdom that evolved from the protists was the kingdom
    a. Fungi.
    b. Plantae.
    c. Animalia.
    d. All of the above
    Answer: D   Difficulty: I   Section: 2   Objective: 3

72. Multicellularity
    a. is a relatively recent evolutionary step.
    b. means that an organism is made up of more than one cell.
    c. is found in some protists.
    d. All of the above
    Answer: D   Difficulty: I   Section: 2   Objective: 3

TEST ITEM LISTING, continued

73. All of the major phyla of animals on Earth today evolved
    a. near the Burgess Shale.
    b. after the five mass extinctions took place.
    c. during the Cambrian period.
    d. prior to the appearance of protists.

    Answer: C   Difficulty: I   Section: 2   Objective: 3

74. The fossil record indicates that Earth has experienced
    a. 5 mass extinctions.   c. 3 mass extinctions.
    b. 4 mass extinctions.   d. 2 mass extinctions.

    Answer: A   Difficulty: I   Section: 2   Objective: 4

75. Two-thirds of all terrestrial life disappeared in the last mass extinction approximately
    a. 440 million years ago.   c. 245 million years ago.
    b. 360 million years ago.   d. 65 million years ago.

    Answer: D   Difficulty: I   Section: 2   Objective: 4

76. A layer of ozone in the atmosphere was critical to the formation of life on land because
    a. land plants need ozone for photosynthesis.
    b. there is a high concentration of ozone in the oceans.
    c. ozone is necessary in order to produce oxygen.
    d. ozone blocks ultraviolet radiation.

    Answer: D   Difficulty: I   Section: 3   Objective: 1

77. The destruction of Earth's ozone layer by industrial chemicals is a valid concern because
    a. animals will not have air to breathe.   c. the climate will change.
    b. harmful ultraviolet light levels will increase.   d. None of the above

    Answer: B   Difficulty: I   Section: 3   Objective: 1

78. Ozone
    a. is composed of three oxygen atoms.
    b. blocks ultraviolet radiation in the upper atmosphere.
    c. made Earth's surface a safe place to live.
    d. All of the above

    Answer: D   Difficulty: I   Section: 3   Objective: 1

79. The associations between the roots of plants and fungi are known as
    a. powdery mildew.   c. lichens.
    b. mycorrhizae.   d. mitochondria.

    Answer: B   Difficulty: I   Section: 3   Objective: 2

80. In mycorrhizae, the fungi provide plants with
    a. food.   c. minerals.
    b. energy.   d. All of the above

    Answer: C   Difficulty: I   Section: 3   Objective: 2

81. The first organisms to populate the surface of the land were
    a. bacteria and plants.   c. plants and fungi.
    b. plants and animals.   d. bacteria and fungi.

    Answer: C   Difficulty: I   Section: 3   Objective: 2

82. While there was no soil present, plants were able to invade the surface of the ancient Earth because they
    a. were supported by insects.
    b. extracted minerals from protists.
    c. could obtain extra nitrogen from the rocky dust.
    d. formed a partnership with fungi.

    Answer: D   Difficulty: I   Section: 3   Objective: 2

TEST ITEM LISTING, *continued*

83. mycorrhizae : plant roots ::
    a. mitochondria : eukaryotic cells
    b. ribosomes : nucleus
    c. mitochondria : nucleus
    d. nucleus : mitochondria
    Answer: A   Difficulty: II   Section: 3   Objective: 2

84. The first animals to invade the land were the
    a. amphibians.
    b. arthropods.
    c. reptiles.
    d. protists.
    Answer: B   Difficulty: I   Section: 3   Objective: 3

85. Insects have been very successful because
    a. they reproduce in large numbers.
    b. they evolved the ability to fly.
    c. some of them can feed on flower nectar.
    d. All of the above
    Answer: D   Difficulty: I   Section: 3   Objective: 3

86. From fossils, we know that insects were the first animals to evolve
    a. pincers on their front legs.
    b. stingers at the end of their tails.
    c. jointed legs.
    d. wings.
    Answer: D   Difficulty: I   Section: 3   Objective: 3

87. Flying insects were able to use the ability to fly to do everything listed below *except*
    a. patrol the entire surface of Earth.
    b. search for food, mates, or nesting sites.
    c. develop a mutualistic relationship with fungi.
    d. transport objects long distances.
    Answer: C   Difficulty: I   Section: 3   Objective: 3

88. Lobsters, insects, and spiders are all examples of
    a. amphibians.
    b. vertebrates.
    c. arthropods.
    d. monerans.
    Answer: C   Difficulty: I   Section: 3   Objective: 3

89. Arthropods have a hard outer skeleton and
    a. a backbone.
    b. hair.
    c. a four-chambered heart.
    d. jointed appendages.
    Answer: D   Difficulty: I   Section: 3   Objective: 3

90. The most diverse group of animals on Earth is the
    a. reptiles.
    b. mammals.
    c. insects.
    d. amphibians.
    Answer: C   Difficulty: I   Section: 3   Objective: 3

91. The ability to fly led to partnerships between insects and
    a. flowering plants.
    b. fungi.
    c. arthropods.
    d. birds.
    Answer: A   Difficulty: I   Section: 3   Objective: 3

92. One major problem faced by organisms moving onto land was
    a. lack of oxygen.
    b. too many competing insects.
    c. body structures that constantly lose water.
    d. All of the above
    Answer: C   Difficulty: I   Section: 3   Objective: 3

TEST ITEM LISTING, continued

93. arthropods : live on land ::
    a. plants : absorb minerals from rocks
    b. fungi : make food
    c. scorpions : fly
    d. insects : fly

    Answer: D
    (were the first animals to)
    Difficulty: II   Section: 3   Objective: 3

94. The first vertebrates
    a. had skeletons made of cartilage.
    b. evolved on the land.
    c. were jawless fishes.
    d. resembled amphibians.

    Answer: C   Difficulty: I   Section: 3   Objective: 4

95. The most successful living vertebrates are
    a. amphibians.
    b. fishes.
    c. insects.
    d. humans.

    Answer: B   Difficulty: I   Section: 3   Objective: 4

96. Vertebrates that are adapted to life both on land and in the water are
    a. reptiles.
    b. arthropods.
    c. bony fish.
    d. amphibians.

    Answer: D   Difficulty: I   Section: 3   Objective: 4

97. Amphibians were able to successfully colonize land because
    a. they can absorb oxygen from the air with lungs.
    b. they have four sturdy limbs.
    c. oxygen-rich blood flows rapidly to the muscles and organs.
    d. All of the above

    Answer: D   Difficulty: I   Section: 3   Objective: 4

98. Reptiles
    a. have watertight skin.
    b. must lay eggs in water.
    c. generally must remain in moist places.
    d. All of the above

    Answer: A   Difficulty: I   Section: 3   Objective: 4

99. Amphibians must lay their eggs in
    a. dry, hot environments.
    b. water or in moist environments.
    c. nests in trees.
    d. winter.

    Answer: B   Difficulty: I   Section: 3   Objective: 4

100. lungs : amphibians ::
    a. bony skeleton : reptiles
    b. watertight skin : reptiles
    c. land animal design : protists
    d. circulatory system : reptiles

    Answer: B   Difficulty: II   Section: 3   Objective: 4

101. many new reptiles : Permian extinction ::
    a. fish and amphibians : Permian extinction
    b. fish : Cretaceous extinction
    c. Permian extinction : beginning of continental drift
    d. mammals, birds, and small reptiles : Cretaceous extinction

    Answer: D   Difficulty: II   Section: 3   Objective: 4

TEST ITEM LISTING, continued

COMPLETION

102. The study of radioactive decay in rocks indicates that Earth is about _____ years old.
    Answer: 4.5 billion   Difficulty: II   Section: 1   Objective: 1

103. Forms of an element that differ in atomic mass are called _____.
    Answer: isotopes   Difficulty: I   Section: 1   Objective: 1

104. Unstable elements that give off energy as they decay, forming stable elements, are called _____.
    Answer: radioisotopes   Difficulty: I   Section: 1   Objective: 1

105. A(n) _____ _____ is the amount of time required for one-half the number of radioactive isotopes in a sample to decay, forming stable elements.
    Answer: half-life   Difficulty: I   Section: 1   Objective: 1

106. By determining the number of half-lives that have passed since the formation of a rock containing radioisotopes and their stable product elements, scientists are able to determine the approximate _____ of the rock.
    Answer: age   Difficulty: I   Section: 1   Objective: 1

107. The _____ _____ _____ is the hypothesis that life developed when molecules of nonliving matter in the ocean reacted chemically during the first 1 billion years of Earth's history.
    Answer: primordial soup model   Difficulty: I   Section: 1   Objective: 2

108. The goal of each of the models that attempt to explain how life's basic chemicals were formed is to explain the development of _____ building blocks.
    Answer: organic   Difficulty: I   Section: 1   Objective: 2

109. Primitive _____ produced the first oxygen in Earth's atmosphere.
    Answer: cyanobacteria   Difficulty: II   Section: 2   Objective: 1

110. Cyanobacteria produced the _____ that is now present in our atmosphere.
    Answer: oxygen   Difficulty: I   Section: 2   Objective: 1

111. The _____ probably were the ancestors of all eukaryotic cells.
    Answer: archaebacteria   Difficulty: II   Section: 2   Objective: 1

112. The theory of _____ is a widely accepted theory that explains the presence of mitochondria and chloroplasts in eukaryotic cells.
    Answer: endosymbiosis   Difficulty: I   Section: 2   Objective: 2

113. Both mitochondria and chloroplasts have their own _____.
    Answer: DNA   Difficulty: I   Section: 2   Objective: 2

114. The first eukaryotes were _____.
    Answer: protists   Difficulty: I   Section: 2   Objective: 3

115. All of the living things on Earth today can be grouped into _____ kingdoms.
    Answer: six   Difficulty: I   Section: 2   Objective: 3

116. Multicellularity allows cells to _____.
    Answer: specialize   Difficulty: I   Section: 2   Objective: 3

117. Having more than one cell is known as _____.
    Answer: multicellularity   Difficulty: I   Section: 2   Objective: 3

TEST ITEM LISTING, continued

118. All of the major phyla on Earth today evolved during the _____ period.
    Answer: Cambrian   Difficulty: II   Section: 2   Objective: 4

119. The death of all members of many different species is called _____ _____.
    Answer: mass extinction   Difficulty: I   Section: 2   Objective: 4

120. At the end of the _____ period 250 million years ago, about 96 percent of all species of animals became extinct.
    Answer: Permian   Difficulty: II   Section: 2   Objective: 4

121. Refer to the illustration above. The three atoms shown make up a molecule of _____.
    Answer: ozone   Difficulty: II   Section: 3   Objective: 1

122. Refer to the illustration above. The molecule shown forms a protective layer in the atmosphere that blocks _____ radiation.
    Answer: ultraviolet   Difficulty: II   Section: 3   Objective: 1

123. Earth's surface is protected from ultraviolet radiation by _____ molecules.
    Answer: ozone   Difficulty: I   Section: 3   Objective: 1

124. The first terrestrial organisms that were able to fly were _____.
    Answer: insects   Difficulty: II   Section: 3   Objective: 3

125. Flying insects form _____ with flowering plants.
    Answer: partnerships   Difficulty: II   Section: 3   Objective: 3

126. The first flying animals were _____.
    Answer: insects   Difficulty: I   Section: 3   Objective: 3

127. The first animals to successfully invade the land from the sea were _____.
    Answer: arthropods   Difficulty: II   Section: 3   Objective: 3

128. Insects were the first animals to evolve _____.
    Answer: wings   Difficulty: I   Section: 3   Objective: 3

129. Insects developed important relationships with _____ plants.
    Answer: flowering   Difficulty: I   Section: 3   Objective: 3

130. Reptiles thrive in dry climates because both their skin and eggs are largely _____.
    Answer: watertight   Difficulty: II   Section: 3   Objective: 4

131. The first vertebrates on land were _____.
    Answer: amphibians   Difficulty: I   Section: 3   Objective: 4

TEST ITEM LISTING, *continued*

132. The large number of marsupials found in Australia and South America can be explained by the movement of Earth's land masses. This movement is known as _____ _____.

    Answer: continental drift   Difficulty: I   Section: 3   Objective: 4

133. Birds and mammals became the dominant vertebrates on land after the _____ extinction.

    Answer: Cretaceous   Difficulty: II   Section: 3   Objective: 4

ESSAY

134. Construct a timeline of events relating to the origin and evolution of life on Earth. Use information presented in your text that represents scientists' current best estimates of when the relevant events occurred. (Your time scale should be in billions of years.) Illustrate your timeline with drawings of cells and organisms that you include. Use descriptions and pictures in your text to help you depict the distinctive features and the environment in which the cells and organisms lived. Include when organic chemicals were first formed, as well as origins for the following: Earth, first cells, first photosynthetic cells, first eukaryotic cells, first multicellular organisms, first animals, first vertebrates, first land plants, first fungi, first land animals, first vertebrates, first mammals, first primates, and first humans.

    Answer:
    Origin of Earth: 4.5 billion years ago
    First formation of organic chemicals: about 4 billion years ago
    - Origin of the first cells: 2.5 billion years ago (These should be depicted as prokaryotic cells.)
    - Origin of the first photosynthetic cells: 3 billion years ago (These should be depicted as prokaryotic cells.)
    - Origin of the first eukaryotic cells: 1.5 billion years ago (These cells should be depicted as containing organelles.)
    - Origin of the first multicellular organisms: 0.63 billion years ago (These should be depicted as small, aquatic animal-like organisms lacking shells.)
    - Origin of the first animals: 0.6 billion years ago (These should be depicted as aquatic animals.)
    - Origin of the first vertebrates: about 0.57 billion years ago (These should be depicted as fishes.)
    - Origin of the first land plants: 0.4 billion years ago (These should be depicted as green and as having roots.)
    - Origin of the first fungi: 0.4 billion years ago (These should be depicted in a color other than green and could be shown growing on rock or on plant roots.)
    - Origin of the first land animals: shortly after 0.4 billion years ago (These should be depicted as arthropods.)
    - Origin of the first land vertebrates: 0.35 billion years ago (These should be depicted as amphibians.)
    - Origin of the first mammals: about 0.25 billion years ago (These should be depicted as small, mouse-like animals.)
    - Origin of the first primates: 0.06 billion years ago (These should be depicted as small, rodent-like animals living in trees.)
    - Origin of the first humans: 0.005 billion years ago (These should be depicted as upright-walking animals.)

    Difficulty: III   Section: 1, 2, 3   Objective: all

TEST ITEM LISTING, continued

135. Most cells found in organisms on Earth today are aerobic; that is, they need oxygen. Explain why the first cells could not have been aerobic bacteria.
    Answer:
    There was no oxygen in the atmosphere of the primitive Earth. Oxygen was found in the atmosphere only after the evolution of the photosynthetic cyanobacteria.
    Difficulty: II   Section: 2   Objective: 1

136. Many scientists think the thinning of Earth's ozone layer is caused by the activities of people in industrialized countries. Why do you think many biologists are urging industrialized nations to take steps that may prevent further destruction of the ozone layer?
    Answer:
    The activities of people might be causing the destruction of the ozone layer. The ozone layer protects us and all living things from harmful ultraviolet light. Its destruction would negatively affect life on Earth.
    Difficulty: III   Section: 3   Objective: 3

137. Why would insects have an ecological advantage over terrestrial animals?
    Answer:
    Flying allows insects to efficiently search for food, mates, and nesting sites. Flight enables insects to disperse easily to new territories, escape predators, and find resources unavailable to terrestrial animals.
    Difficulty: II   Section: 3   Objective: 3

138. What characteristic did the first terrestrial animals have that allowed them to survive on land?
    Answer:
    The first land animals were arthropods, which had hard outer skeletons that prevented them from drying out.
    Difficulty: III   Section: 3   Objective: 3

139. Describe the structural innovations that helped amphibians adapt to life on land.
    Answer:
    A number of adaptations evolved in amphibians that allowed them to live on land. These structural changes included the evolution of lungs and changes in the circulatory system. In addition, amphibians evolved limbs and a skeletal system that allowed the amphibians to walk on land.
    Difficulty: II   Section: 3   Objective: 3